SpringerBriefs in Energy

Energy Analysis

Series editor

Charles A.S. Hall, Syracuse, USA

More information about this series at http://www.springer.com/series/10041

S.W. Carmalt

The Economics of Oil

A Primer Including Geology, Energy, Economics, Politics

Springer

S.W. Carmalt
SW Consulting SA
Chambésy
Switzerland

and

Department of Earth Sciences
University of Geneva
Geneva
Switzerland

ISSN 2191-5520 ISSN 2191-5539 (electronic)
SpringerBriefs in Energy
ISSN 2191-7876
SpringerBriefs in Energy Analysis
ISBN 978-3-319-47817-3 ISBN 978-3-319-47819-7 (eBook)
DOI 10.1007/978-3-319-47819-7

Library of Congress Control Number: 2016954908

Printed on acid-free paper

This Springer imprint is published by Springer Nature
The registered company is Springer International Publishing AG
The registered company address is: Gewerbestrasse 11, 6330 Cham, Switzerland

Preface

In the early 1970s I was a graduate student in geology. And I was also sitting in gas lines due to the Arab Oil embargo. When I was offered a job with Cities Service Company, still a major oil company, I thought I would be able to both do good for my fellow citizens by finding more oil and collect a generous salary while doing geology.

When I started work, I quickly realized that the oil business was a great deal more than just finding oil. That, of course, was and is necessary. But there was the small matter of how the company would show a profit from this activity. After I had spent several years evaluating the geology of potential exploration targets all around the world, and perhaps because I was one of the few geologists who had a rudimentary knowledge of computer programming, I was asked to help do the calculations to determine whether the ventures would be profitable.

As my career developed, this led to positions in the company's planning department, looking at strategic issues. The oil price increases of the 1970s had the entire industry talking about M. King Hubbert's theory that oil production would peak, and I was assigned to study this and other long-term issues.

Cities Service Company disappeared in the consolidation of the oil industry that took place in the first part of the 1980s. And although I had tried to avoid it, I found myself unemployed. So were most other geologists, so I worked as a computer specialist for a number of years.

But things do sometimes move in circles and beginning in the late 1990s oil shortages were again on the horizon. As it happened, my computer consulting work landed me in an environmental NGO which was concerned about climate change, sustainability and other environmental issues on a global basis. Many environmentalists believed that global peak oil would save the climate—and the world. Although political agreement seemed impossible, the oil would simply "run out" and earth would be saved. My previous work in the oil industry made me skeptical that it would work out this way.

Kenneth Deffeyes, who might be called a disciple of M. King Hubbert, predicted that the global oil production peak would be in November of 2005. He was both correct and wrong: correct because production from the type of oil deposits that both Hubbert and Deffeyes discussed did start to decline in 2005, but wrong because there are other types of oil deposits. As was the case when I started working for the oil company, there is a great deal more involved than just the way oil has accumulated in geologic formations.

This interplay between geologic oil accumulations, pure economics, and the impact that political processes and events have on the oil economy continues to fascinate me. We all want to know about the future; perhaps even predict it. The challenge is to make sense out of a great many anastomosing trends. It is a challenge—and it is great fun.

My feeling is that the world we have known since the industrial revolution is changing. Indeed, it must change due to population pressure, resource availability, information technology, and climate change—and probably other factors. In this book my aim is to look at the oil industry and how it is being affected by all these changes. It has been a fun book to write, and I hope you will enjoy it.

Along the way, I have had the benefit of a great many stimulating conversations. The list is long, but I do want to give special mention to: Roger Bentley, Arthur Dahl, Ken Deffeyes, John Gault, Joachim Monkelbaan, Andrea Moscariello, Ken Russell, and Deborah and Frank Vorhies. In some cases they may disagree strongly with what I have said, so I want to be clear that they are not responsible for the ideas presented here. But that does not detract from either their intellectual stimulation nor, more importantly, their friendship.

And final thanks to Charles Hall, the series editor. In the course of writing he has been patient and very helpful.

Chambésy, Switzerland S.W. Carmalt
November 2016

Contents

Chapter 1
Introduction

Our economy runs principally on fossil fuel energy: coal, oil, and natural gas. Very roughly each provides about 30% of the world's energy, with the final 10% coming from all other sources—nuclear, hydro, wind, solar thermal, photovoltaics, geothermal, biomass, and so forth. The 30% of oil energy is of particular importance because oil is the primary power source for transportation; the economy depends on transportation to move raw materials, finished goods, people, and even information from one place to another. If we run out of oil the global economy is in serious trouble.

Providing enough oil for the economy is a commercial undertaking, albeit with major government involvement over the years. The finding, producing,[1] refining, and selling of oil is a major global industry, dependent on the massive financial investments required to support all aspects of the petroleum industry. The oil price is part of the daily financial news, affecting everyone from the chairman of ExxonMobil to a villager in the Mekong Delta. Oil companies, whether publicly owned by many shareholders, state owned, or privately held, invest billions of dollars each year in keeping the economy fuelled and lubricated; a large number of specialized service companies support this operation, providing combined employment for millions of people.

Government involvement in the oil industry is significant. While the image of "big oil" generally is of companies such as ExxonMobil or Chevron, which are publicly owned by shareholders, approximately three-quarters of world oil production is actually under the control of companies whose majority owner, if not sole shareholder, is a national government (Helman 2014). Major oil-producing countries such as Saudi Arabia, Norway, Nigeria, Russia, and Venezuela derive much of their national income from the production of oil. For many other countries, the USA is an example, the cost of importing oil is a major component of their foreign

[1]"Producing" is the term generally used in the oil industry. Others prefer the terms "extracting" or "exploiting."

S.W. Carmalt, *The Economics of Oil*, SpringerBriefs in Energy,
DOI 10.1007/978-3-319-47819-7_1

exchange balance. Thus oil trade arrangements and the price of oil become important factors both for national policy and in international relationships.

Oil is bulky and storage is thus expensive; this means that disruptions in oil supply at any point between the oil well and the final user can cause economic havoc. The politically-based oil supply disruptions of the 1970s influenced between 20 and 25% of the global supply for only a few months, but the result was a global pause in economic growth (BP Statistics 2014). From 1961 to 1973 world GDP growth was over 4.0% every year but fell to close to 0% in 1974 and 1975; after recovering in the late 1970s, growth again plummeted in 1980 when the fall of the Shah of Iran caused a second supply disruption.[2] In each of these two instances, the USA was actually pushed into economic recession. Government involvement extends beyond just the production of oil; taxes, subsidies, and regulations at every stage of the industry have a major influence, generally designed to ensure a reliable supply of the energy provided by oil.

Oil economics means various things to various people. For some it is primarily a question of whether a particular company will be profitable over the coming quarters or years; or the concern may be what the impact of gasoline pump prices will be on the overall economy. For others, oil economics may represent the power that the oil industry has within the political process. Still others may see oil economics in the context of the environment, where it provides the current framework against which alternative energy supplies must be measured. Strategic planners in all areas, including the oil industry, ponder the continued availability of this valuable resource on a finite earth.

Oil economics is tightly linked to natural gas economics. From the geology in which oil and gas are formed, through the drilling methods used to find and extract them, to the engines and turbines that convert their energy into the form in which we use it, these two fuels are closely linked. While the focus here is on oil, the overlap of the two energy sources in both technologies and companies means that this book will often encompass the economics of natural gas as well. We will start with an examination of the immediate profitability of oil extraction and then broaden our outlook to cover many other aspects of oil economics.

---

[2]GDP statistics taken from the World Bank online database (http://databank.worldbank.org accessed: 2014-12-17).

Chapter 2
Oil Company Finances

It is really very simple. Oil and gas companies are in business to make money for their shareholders—individual shareholders, or in the case of state-owned companies, the owning government. Making money means that revenue is greater than expense. The devil is in the details.

2.1 How Oil Companies Make Investment Decisions

To better understand how oil companies make their decisions, we will look at a very oversimplified financial example for drilling a small oil prospect. An oil company is in the business of finding, producing, and selling oil and its refined products. Some companies do only one or two of these things, whereas the large integrated oil companies are involved from the initial discovery well to the sale of gasoline to a final customer. Indeed, a similar sort of financial model is used for almost every investment decision made by any corporation within a market economy. While this basic model is developed for shareholder-owned private oil companies, the economic factors are essentially the same for state-owned oil companies as well.

An oil exploration company will always be on the lookout for good places to drill. Suggestions come from all sorts of places; historically a natural oil seep might suggest a good location; other oil wells certainly indicate an interesting area, although whether there is an available location for additional well(s) is another matter; property owners, ranging from single homeowners to national governments, may suggest that they have a good prospective area; other oil companies may offer to share the risks of exploring a specific location; and an oil company of any size will have employees whose job is simply to suggest locations. The history of the oil industry is full of tales, sometimes humorous, about when and why various people decided to drill where they did.

S.W. Carmalt, *The Economics of Oil*, SpringerBriefs in Energy,
DOI 10.1007/978-3-319-47819-7_2

Once an exploration target is identified, the oil company has a prospect. Our simple prospect will be small—to better understand the finances we will assume that only one well will suffice to produce all of the oil (this is unrealistic). With our prospect identified, we now have to begin investing some money. One of the most important features of the oil business is that large amounts of money need to be spent before we have anything to sell as a result. As the work goes along, the company will always be asking the question "is this going to make us money?" Many projects are abandoned at various points along the way. It is worth noting that the profits from the successful prospects need to be sufficient to cover the expenses of those that are not successful.

2.2 Who Owns the Resource?

The first requirement is to make sure that this is a location in which a well can be drilled, and that our company has the legal right to do so and potentially profit from a success. In most of the world, minerals and resources at depth belong to the government; in the United States (and a few other places) the land surface owner is presumed to own everything beneath the land. Either way, the owner of the resource will want some compensation, a royalty,[1] for the oil and gas produced. The oil company employees who negotiate with the landowners are called landmen in the industry. In the USA and other places of private mineral ownership landmen visit property owners with offers to either lease or purchase the mineral rights. They spend time in county courthouses researching property ownership records, and talking with landowners or anyone else who can help them find the owners of the mineral rights. The company is generally interested in the minerals now, so the standard starting point for the negotiation is to lease the mineral rights in return for a fee called a "signature bonus" plus the future royalty. There are as many variations in lease terms as one can imagine; price per acre for the signature bonus, percentage of revenue from the well, and so forth. Because the mineral rights are private property, they can be bought and sold separately from the surface property, causing complex ownership problems. The landmen have to sort all this out before drilling. In this context J. Paul Getty is reported to have quipped, "the meek may inherit the earth, but not its mineral rights".

Outside the USA the situation is much different; generally the government owns the mineral rights and landmen might better be called "commercial diplomats," negotiating directly with government officials. The government will typically create fairly large areas, often called blocks, which are, in effect, leased to an oil company.

[1]The term royalty goes back to when government ownership of the land meant it was owned by the king.

Again, terms of the lease can be almost whatever one can imagine—I once heard of an oil company agreeing to invest in a bicycle manufacturing operation as part of a lease negotiation.

While the landmen are working on the ownership terms, the company will continue to study the area to refine the assessment of the oil likely to be present. From regional geologic studies, from data that is bought and sold within the oil industry, and by using various models (Moscariello 2016), the company will examine everything it can about this prospect. The most important data are generally seismic surveys that have been done.

2.3 Seismic Surveys: Assessing the Resource

The basic principle of a seismic survey is to create vibrations at one point and then see how long they take to be reflected back to another point as shown in Fig. 2.1.

When oil companies first started doing seismic surveys they used small dynamite explosions for the seismic source; today they use vibrating trucks on land or air guns at sea to produce the seismic signal. Moving the source and geophone collectors along a line, and then processing the results using computing power, results in a type of cross section of the rocks underneath. An interpreted version of such a seismic section is seen in Fig. 2.2. In this example, the sedimentary layers of the Karoo rocks can be seen as having been tilted and folded; these sedimentary beds have also been offset by faults. Beneath the sedimentary layers one can see some nonlayered "basement" rock, which is generally not thought to be of interest for petroleum.

Seismic processing uses the biggest and fastest computers, and today instead of simply progressing along a line, the geophones are frequently arranged in a grid so the resulting processing can provide a 3-dimensional view of the rocks and likely oil or gas accumulations.

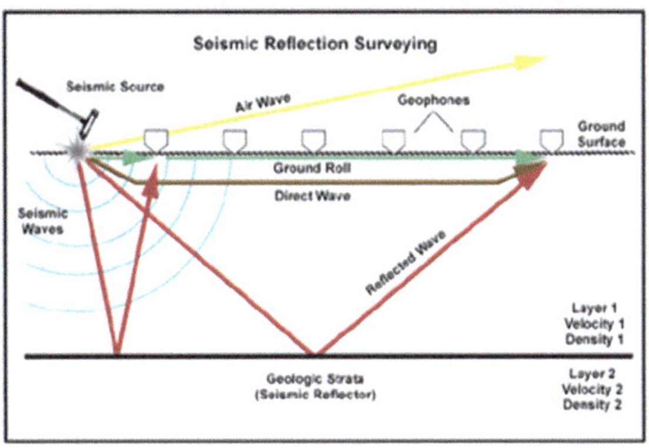

Fig. 2.1 Seismic Reflection Methodology [*Source* Illinois State Geological Survey (2012)]

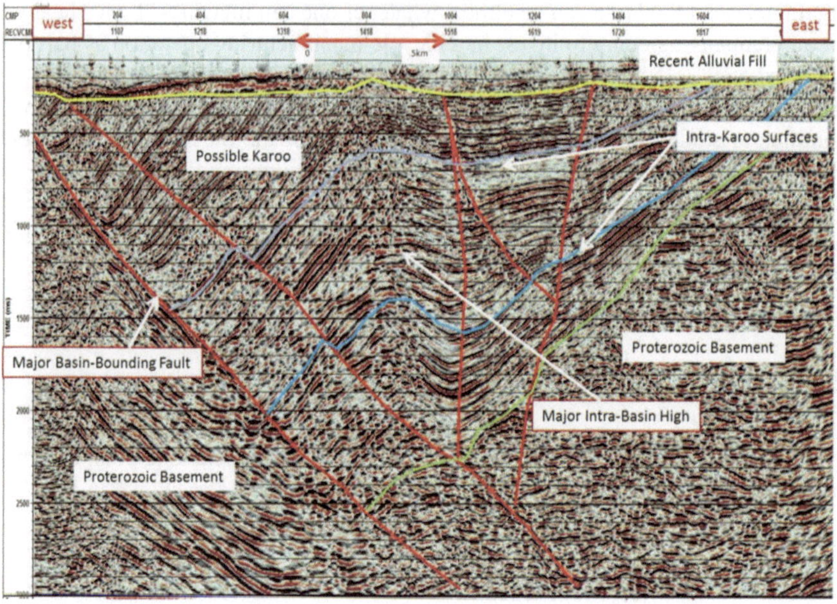

Fig. 2.2 An interpreted seismic line [*Source* Swala Energy, PennEnergy (2013)]. The "Karoo" is a group of layered sedimentary rocks. The *black* and *dark red lines* represent geological strata, the *bright red lines* faults; there might be oil trapped in the folds at the top of the Major Intra-Basin High

2.4 Drilling a Well

Presuming that our analysis so far indicates that we should drill a well, our oil company decides to go ahead. This is the big expense! Our simple prospect will be on land; there are additional considerations for offshore operations that make them even more complex and expensive. Civil engineers will prepare the site for all the facilities needed to support the actual drilling. Specifically, the site needs to be accessible, which may require building an access road; the working area will need to be leveled and well drained; either lined holding ponds or storage tanks will need to be available for the drilling mud; and provisions made for fuel storage, electricity, water, and sewage facilities for crews. The area will need to be fenced and gated to keep wildlife out and for security. Usually an oil or gas company will contract each aspect of the operation to a firm that specializes in that particular task. A contract will be given to a specialist drilling company, which will provide the drill rig and crew to do the actual drilling; subsidiary contracts will be given to companies to monitor the well's progress, provide safety equipment, do specific tasks such as cementing casing, engineer the drilling mud, do geophysical logging, and whatever else is needed to ensure, as best as possible, that all will go to plan.

The drilling itself starts when the drill bit first enters the ground. This is the spud date, which is one of the statistics frequently reported and used for analysis. After drilling has progressed into solid rock, a large diameter pipe called casing will be fitted into the hole and any space between the outside of this pipe and the rock is filled with cement. When this is done properly it ensures that the inside of the well and the rock through which it has been drilled remain isolated from each other. Particularly at shallow depths this isolation is important; it protects fresh water aquifers. The drilling then progresses at a slightly smaller diameter. As the well gets deeper, casing may be used repeatedly; there is an obvious trade-off between necessary isolation of the well from the surrounding rock and the fact that each time a new casing is installed it makes the well smaller. Figure 2.3 illustrates a typical casing program for gas prospects in the Marcellus shale of western Pennsylvania.

When wells have problems, one of the most frequent causes is some sort of problem with the cement between the casing and the surrounding rock.

During the drilling process, the drill bit turns against the rock at the bottom of the well, breaking up the rock. The drill is at the bottom of a heavy steel pipe, the "drill stem." A heavy fluid called mud is pumped down the inside of the drill stem; it both cools the drill bit as it cuts and circulates back to the surface between the drill stem and the inside of the casing, carrying chips of the drilled rock with it. The mud is a carefully engineered clay slurry; both the physical and chemical properties can be critical, as the properties of the rocks through which the well is being drilled will also vary. The rock chips, called cuttings, are continuously returned to the surface by the circulating mud and are examined. Up to this point, all the information at this particular location has been gathered indirectly, but the cuttings are direct samples of the geology. The information gathered is plotted against the depth, with the result being the well log. Other measurements, for example electrical properties, are also plotted against depth, providing electric logs. These used to require periodic interruption of the drilling so that specialized instruments could be lowered into the well on a cable, with the readings being recorded as "wire-line logs"; today this type of wire-line log may still be used, but many of the measurements now can be made in real time with instruments mounted close to the drill bit and results telemetered to the surface in real time.

When the target depth and rock formation has been reached, it is time to test the well. Almost everything depends on these first tests, which will determine whether the well will produce sufficient oil or gas for it to be an economic success. The tests will measure the rate of production over a period of time. Rates for producing oil wells can be between a few tens of barrels per day up to over 10,000 barrels per day. What is considered "good" will depend on specifics of the individual prospect, e.g., well depth, onshore or offshore location, infrastructure to move the oil to market, and many other factors. Gas wells can produce from a few tens of thousand cubic feet per day to over 50,000,000 cubic feet per day (cfd). Presuming the tests are satisfactory, the next step will be to prepare the well for its life as a producing well; if not, the well will be sealed off with concrete and considered plugged and abandoned.

Fig. 2.3 Casing program in the Marcellus area [*Source* Frantz (2014)]. The vertical scale has been greatly compressed, as oil wells are typically more than 1500 m (5000 ft) deep

A short digression: How to measure oil and gas

Oil is generally measured in barrels. This is a volume, defined as equal to 42 US gallons. The use of 42-gallon barrels dates back to the mid-nineteenth century beginnings of the US oil industry in Pennsylvania. Over the years, most other producing areas have adopted this standard measure. The notable exception is that the former Soviet Union measured oil by weight, and countries that were within that economic area still frequently report oil production in metric tons rather than barrels.

Conversion between barrels and metric tons is not straightforward. Different crude oils have a range of 6.5 barrels per metric ton (the heaviest oils) to 7.9 barrels per metric ton (the lightest oils). A value of 7.33 barrels per metric ton is generally used if a value for the specific oil is not known.

Natural gas is measured by volume. In the USA this is cubic feet; thousands of cubic feet are generally used (confusingly abbreviated mcf or MCF). In many other countries the measurement is in cubic meters, although the influence of the major US companies will sometimes result in cubic feet being used even outside the USA. The conversion is 1 m^3 = 35.315 cubic feet (or 1 cubic foot = 0.0283 m^3). Gas is most often bought and sold by its energy content, which will vary somewhat depending on the specific gas source. It happens that one thousand cubic feet of methane (1 MCF) contains approximately one million BTUs (BTU stands for British Thermal Unit, still used instead of calories or joules or kilowatt-hours as an energy measure in the USA). Quoted gas prices are frequently for millions of BTUs (conveniently thousands of cubic feet). North America has the most developed natural gas markets, and a price of $3.85 per million BTUs is thus approximately $3.85 per MCF.

One barrel of oil has the approximate energy equivalent of 5.8 MCF of natural gas. Because both gas and oil can vary in their energy content, using 5.8 for gas-to-oil conversions is yet another approximation. Generally an even more approximate ratio of 6 MCF per barrel is used to convert natural gas to oil equivalents when making regional or global comparisons or summations. Comparisons of either oil or natural gas to other energy sources, such as coal or hydroelectric power, have yet more methodological details to consider. Generally the approach is to convert everything to either barrels of oil (the converted amounts become barrels of oil equivalent or boe) or to kilowatt-hours (kwh). Using the SI energy unit of joules is officially recommended. The units in such comparisons are frequently millions, billions, or trillions and care must be taken with this aspect of any conversion.

2.5 How Much Will All This Cost? Will the Company Make Money?

The basic principles of financing our oil prospect are those of project financing. We can see an oversimplified example of a financial model of our project in Table 2.1. The "Profit after tax" is what many would call the "bottom line."

While oversimplified to the point of being unrealistic, Table 2.1 is a useful example to illustrate some important points in oil company decision-making. Before we start to analyze our financial model, here are a few more details used in this specific example

We estimate that the production will decline 12.5% each year from the previous year's value. This is not realistic, but will allow our model to illustrate some points. More will be said about decline rates later. The oil price will be $90/bbl throughout our project; this is much higher than the current oil price. Putting in $40/bbl will make our project uneconomic, but we may keep this in our file and activate the plan if we think prices will return to higher levels in another year or two. The landowner will receive a royalty of 15%. Because of the time to develop the prospect, prepare the site for drilling, and drill the well, we will only see the initial production in year 3. From year 4 the well will produce for 300 days each year; this allows 2 days per month for any maintenance operations, etc.; year 3 will have only a partial year of production. Initially the well will produce naturally as a result of subsurface pressure, but in year 5 we decide that we have to spend an additional $1 million to install a pump because the pressure is declining. Unrealistically, the production decline curve and days without production in our model are unaffected by this. The operating costs include a flat charge of $500,000 per year to cover the costs of the corporate office (executives, lawyers, accountants, etc.). In addition, costs directly attributed to this particular producing well will be 10% of the revenue from the well. This governmental jurisdiction has a 10% tax (severance tax), based on the value of oil produced; this is essentially an additional royalty. We will presume that the company is paying a 20% income tax on its profits.

These are very simple assumptions, but they allow us to examine the financial model and call attention to some of the factors that will be considered in making a decision as to whether to drill this well. A first look at this financial model shows that the company will make $10.2 million from this well before income tax, and $6.7 million after tax. The after tax number is a guess, because the profits from this project will be combined with profits and losses from other corporate activities to determine the overall company tax.

Figure 2.4 shows the cumulative cash flow for this well.

As we have already noted, the costs are skewed to the beginning of the project. Our company will need to have $8 million available to invest in this, and it will be only in year 7 that this money will be recovered. If this money is borrowed,

Table 2.1 A simple economic model for a one-well oil field

	Year 1	Year 2	Year 3	Year 4	Year 5	Year 6	Year 7	Year 8	Year 9	Year 10	Year 11	Year 12	Year 13	Year 14	Year 15	Totals
Revenue																
Oil (bbls per day)			250	219	192	167	147	128	112	98	86	75	66	58	50	
Price/bbl	90	90	90	90	90	90	90	90	90	90	90	90	90	90	90	
Sales revenue	0	0	21,60,000	59,13,000	51,84,000	45,09,000	39,69,000	34,56,000	30,24,000	26,46,000	23,22,000	20,25,000	17,82,000	15,66,000	13,50,000	3,99,06,000
Expenses																
Predrilling	5,00,000	10,00,000														15,00,000
Drilling		50,00,000														50,00,000
Completion			7,50,000		10,00,000											17,50,000
Royalty			3,24,000	8,86,950	7,77,600	6,76,350	5,95,350	5,18,400	4,53,600	3,96,900	3,48,300	3,03,750	2,67,300	2,34,900	2,02,500	59,85,900
Operating costs	5,00,000	5,00,000	7,16,000	10,91,300	10,18,400	9,50,900	8,96,900	8,45,600	8,02,400	7,64,600	7,32,200	7,02,500	6,78,200	6,56,600	6,35,000	1,14,90,600
Severence tax			2,16,000	5,91,300	5,81,400	4,50,900	3,96,900	3,45,600	3,02,400	2,64,600	2,32,200	2,02,500	1,78,200	1,56,600	1,35,000	39,90,600
Total expenses	10,00,000	65,00,000	20,06,000	25,69,550	33,14,400	20,78,150	18,89,150	17,09,600	15,58,400	14,26,100	13,12,700	12,08,750	11,23,700	10,48,100	9,72,500	2,97,17,100
Results																
Profit before tax	−10,00,000	−65,00,000	1,54,000	33,43,450	18,69,600	24,30,850	20,79,850	17,46,400	14,65,600	12,19,900	10,09,300	8,16,250	6,58,300	5,17,900	3,77,500	1,01,88,900
Taxes (20% of profit)	0	0	30,800	6,68,690	3,73,920	4,86,170	4,15,970	3,49,280	2,93,120	2,43,980	2,01,860	1,63,250	1,31,660	1,03,580	75,500	35,37,780
Profit after tax	−10,00,000	−65,00,000	1,23,200	26,74,760	14,95,680	19,44,680	16,63,880	13,97,120	11,72,480	9,75,920	8,07,440	6,53,000	5,26,640	4,14,320	3,02,000	66,51,120
Metrics																
NPV @ 3%	−10,00,000	−63,10,680	1,16,128	24,47,784	13,28,892	16,77,498	13,93,473	11,35,986	9,25,567	7,47,961	6,00,811	4,71,741	3,69,375	2,82,132	1,99,658	43,86,327
IRR	13.0%															

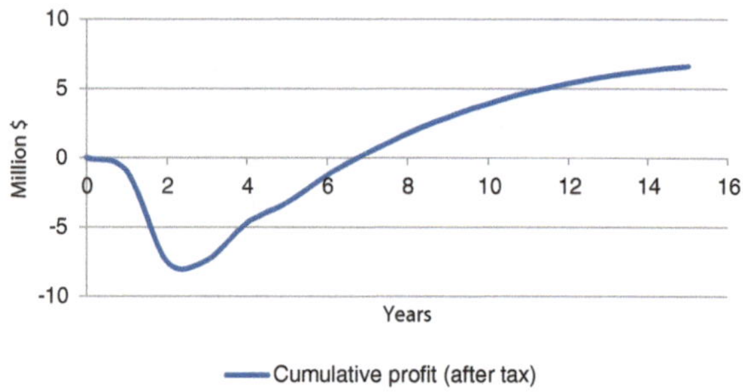

Fig. 2.4 Cumulative profit (cash flow) for a simple project

we would have to pay interest on it; normally a company will include the costs of such investment funds in its financial models to reflect the cost of using money for this particular project.

In addition to the cost of the initial period of negative cash, the time to payout, the time when the company will have at least recovered the money spent, will also be a factor when the company is putting together a portfolio of projects.

Perhaps the first thing that one notices in our financial model is that the price of oil has been kept constant over our project. We are clearly going to have to use an educated guess about this, but it is obvious that whatever this educated guess is, it will be important in the financial analysis. For this reason, oil companies generally adopt a company-wide price forecast that is used in all project evaluations. This allows managers who have to select among projects to have a consistent basis for comparison; the larger oil companies have entire offices whose primary deliverable is the company price forecast.

In addition to having to estimate the oil price into the future, the rate of production also will be an estimation. If our well is being drilled in an area with existing oil production, the initial amounts and decline percentages may be fairly well known. But in less developed areas there may be more uncertainty in these estimations. Costs also have to be estimated into the future. In short, preparing just this simple model will have required the expertise of a number of specialists; they will have used their professional judgement and made their best estimates. There will always be details that are incorrect; the basic question is how incorrect?

Our simple financial model will form the foundation for deciding whether to drill this well. As such, it will be compared with models for other prospects and projects. How best to compare them? The numbers themselves are important, but some calculations will make the job easier. These metrics are an important part of the process.

2.6 Financial Metrics

The first metric that we will consider is the Return on Investment (ROI). We are discussing finances here, so, strictly speaking, this is the Monetary Return on Investment.[2] But when used in a financial context, which is the most frequent use, ROI is the monetary or financial return on investment. ROI is a ratio, and the financial ROI is defined as:

$$\mathrm{ROI} = \frac{\text{Gain from investment} - \text{Cost of investment}}{\text{Cost of investment}}$$

In our example, we consider the Gain as being the sales proceeds less the operating expenses, and the Cost of investments as the predrilling, drilling and completion costs

$$\mathrm{ROI} = \frac{\$18,438,900 - \$8,250,000}{\$8,250,000} = 124\%.$$

In the above calculation we have ignored the notional income tax that the company will pay on the project returns. If we include this in the calculation, the we get

$$\mathrm{ROI} = \frac{\$14,901,120 - \$8,250,000}{\$8,250,000} = 81\%.$$

This illustrates one of the major problems with ROI analysis—the result can differ depending on what, exactly, is included in the calculation. While ROI is frequently used to compare projects and companies, care must be taken to ensure that each calculation is made in a similar manner. For example, in our simple project we have considered the “pre-drilling” expenses as a capital (investment) item. But are these expenses really investments if they are made before we take the decision to drill the well? Deciding questions such as this is the daily work of the accounting and tax departments (which may use somewhat different definitions and come to different answers). When using the ROI to compare one project to another, both within or between companies, the first consideration needs to be whether the calculation was done in the same way for each.

A second major issue with ROI calculations is that they do not account for timing. The after tax calculation above gives us 81% after 15 years. But if we only look at 8 years, then the same calculation is

[2]In Chap. 6 we will discuss at some length an energy equivalent calculation in which all values are in energy units. When calculated with energy units the abbreviation always includes an E, EROI, EROEI and EROIE are all used.

$$\text{ROI} = \frac{\$10{,}049{,}320 - \$8{,}250{,}000}{\$8{,}250{,}000} = 22\%.$$

Which figure should we use?

The calculation of ROI highlights an important issue. The "Cost of investment" or capital expenses are not the only expenses in the project—there are also the operating costs, which have to be deducted in order to arrive at the "Gains from investment." The latter costs are not considered an investment, and hence not included in the denominator of the ROI calculation. These two categories of disbursement are frequently shortened in discussions to "capex" and "opex" for capital expenses and operating expenses. Not only are they treated differently in the ROI calculation, but also they are treated differently in a number of other accounting summaries, and in the tax treatment they receive. This division of expenditures can be discerned by analysis of the financial reports that public companies are required to publish. Conceptually, capex represents the money that the company is investing in its future, whereas opex is the money that it has to spend just to stay in business. While there are some differences in the way the publicly traded oil and gas companies classify their expenses, the capital investment figures are useful in comparing companies. If a company's capex falls too low, it means that it is not investing in the future. For this reason, capex comparisons are frequently made when choosing which oil companies will be better investments. Within an oil company, it may be important to understand how much capex is being invested in exploration as compared to how much is invested in refining operations; within the oil and gas industry, one might compare capex in oil exploration with capex in gas exploration; and within society it may be useful to understand the capex put into fossil fuels compared to that put into renewable energy sources.

Because ROI ignores the time dimension, how best should this element be incorporated into our project metrics? We have already noted that most oil and gas projects have long project lives and that the cost of invested money is included in financial projections. Two calculations are frequently used to include time considerations: Net Present Value (NPV) and the Internal Rate of Return (IRR). Both are available as formulae in spreadsheets, making them easy to calculate.

The NPV calculation converts all future cash flows to the present at a specified interest rate. By subjecting all such cash flows to these interest calculations, a single "present value" of the project is determined. One way such a calculation could be useful is if we have decided to go forward with our prospect, but before spending any money on it, another company comes and asks to buy the prospect. What should we sell it for? The NPV would be a starting point for such a negotiation because it represents the value today if all the future revenue were available to be invested at the specified interest rate.[3] When using the NPV to compare projects, it

[3] In this situation the rate would be called the discount rate rather than the interest rate. It is the same thing.

is important that the interest rate used in the calculation be the same; hence an NPV should always specify the interest rate used (e.g., NPV at 3%).

Related to the NPV is the IRR, which is simply the interest rate that results in an NPV of 0. This allows different projects to be compared directly with a single number. If we are evaluating two prospects, one with an IRR of 9% and another with an IRR of 15%, the latter is financially the better project. What IRR does not do, however, is take into account the size of the project, so we do not know whether the 15% project will make a meaningful difference in the amount of oil the company has available to sell in 10 years, or will, perhaps, bankrupt the company before production starts—either extreme is possible. Yet in both cases the individual prospect may make good or poor financial sense. A clear presentation of some of these metrics is provided by Henriksen (2004).

All of the metrics for the finances that include the time value of money have one thing in common—the lower the interest rate, the more impact the results many years in the future have on the present evaluation. To put this another way, when interest rates are high, the value of a distant revenue stream is low. Thus, during times of high interest rates projects will skew toward rapid returns. Conversely, when interest rates are low, the time value of money is low and long-term projects become relatively more attractive. Given that many oil and gas industry projects have 30–50 year life expectancies, this is important, although seldom would a one-well prospect be quite so long-lived.

2.7 More Complex Projects

An offshore project may take up to 10 years between the initial decision to go forward and the time of first production. Assuming the first well is a success, several additional appraisal wells may need to be drilled before enough is known about the field to make good decisions about how to produce it. After that, platforms have to be designed, constructed and installed. Then producing wells have to be drilled and completed. Finally, the infrastructure needed to move the oil or gas from the offshore facility to a buyer must be constructed. These projects frequently cost billions of dollars, so our simple one-well spreadsheet is only just the beginning.

But the principles remain the same. It is just that the spreadsheet(s) become much larger and more complex. In practice, specialized industry computer models are used to calculate the projected financial returns and metrics for a prospect.

Developing major gas fields, both onshore or offshore, is typically an equally long-term endeavor. The issue with gas is that it is more difficult to transport than oil. For any significant quantity, either pipelines need to be built or the gas needs to be liquefied. Natural gas is primarily methane, which becomes a liquid when cooled below −162 °C. Such a refrigeration plant, when required, is both a major capital expense in the construction budget, and operating it will consume 8% to 15% of the gas originally produced (Foss 2007; Chandra 2014). Again, doing the engineering design and arranging for construction will consume considerable time and money before the project can be brought "on stream" and the first revenue arrives.

Throughout the development and execution of a prospect the project will always be under financial review. The question is: "are things working out financially?" At some points in the life of the project the question will be given especially close scrutiny. These points will depend, in part, on the amount already invested and the nature of the prospect. For example, in many countries the contract that gives permission to explore and produce oil or gas may require that a certain number of wells be drilled; once such a contract is signed, the question of whether to go ahead with drilling a well takes on a different nature. But, in general, the critical decision points are: whether to acquire the rights and proceed with the project, whether to proceed with drilling, whether to consider the well a success and complete it for production, and at a later time, whether to install a production-enhancing technology (a pump in our simple example), and finally, when to plug and abandon the well.

All business decisions are made with imperfect knowledge. Two of the major risks in our simple oil prospect model are: that the future oil price is not what we have used, and that we do indeed find oil and produce it at the rate projected. There are many factors that affect the future price of oil; experienced, knowledgeable analysts may disagree markedly. In 2014 the commodities group at Citicorp, the global financial institution, projected that oil prices would drop into the $70 per barrel range while concurrently the CEO of Chevron was saying that "$110 per barrel oil is the new normal" (Kopits 2014). The only truth is that that these two oil price forecasts could not both be correct. We will come back to this issue as we examine the global context of oil in subsequent chapters.

The other major risk is that we will not find the oil we expect. Typical oil fields are at depths of between 1500 and 4000 m (ca. 5000–12,500 ft). While modern seismic surveys, especially 3-D detailed predrilling surveys, can convey a great deal of information about the subsurface, it is only when the prospect has been drilled and tested that we really know the composition of the fluid and rock characteristics. Geologic textbooks notwithstanding, the subsurface strata are not uniform layers of rock; rather, rocks change in their detailed nature both laterally and vertically at a scale of centimeters and meters. So the indirect measurements made from thousands of meters above will naturally be imperfect.

Of course, oil companies do not discover new deposits of oil with "one well" projects. A single-well analysis would be done only as part of developing a larger project—an oil field. But the financial analysis for a multi-well oil field is essentially the same, just more complex. The costs of drilling the wells are all added together, as is the revenue from all the wells, with the calculations being done on the totals. Depending on the geographical remoteness of the oil prospect, the infrastructure costs can easily become significant. This is particularly true when operating offshore, where it will probably be necessary to construct platforms which can easily run into the hundreds of millions of dollars.

2.8 Government Policy Impacts

Before we move on to more than just our single-well prospect, there is one more important aspect that needs to be added to our financial model: government policy. As an example, we will use the government's tax policy with respect to capital investment depreciation. Astute readers will note that Table 2.1 calculates the company tax on the year-by-year results of the cash flow, with no tax if the result is negative and a tax if the result is positive. But capital expenses are generally depreciated; rather than the spending being included in the results of the year it is spent, the amount is spread out over the life of the project. How to spread it out is the province of tax regulations and the way that the company's tax specialists interpret them. To keep our example simple, we will presume that the government where we drill our well taxed entirely on the basis of cash flow at one time (the example already given), but then changed to allow capital expenses to be depreciated in proportion to the amount of oil produced. The revised project finances for this new policy is shown in Table 2.2. What is important to note is that the additional complexity of Table 2.2 relates only to the accounting treatment of various expenses—the real amounts spent and received are the same. These changes, which change only the timing of the income and expenses in the model, result in significantly different project metrics, without changing the cash spent (except for tax payments) or received. Simply by changing the deprecation rules for capital investment in our model the IRR increases from 13% to 78%; the NPV at 3% increases from $4.4 million to $6.5 million and the simple cash result from $6.7 million to $7.9 million.

There are two underlying reasons for the significant changes in our model. The first is that our simple tax calculation is done year-by-year. When we were investing most of the capital, in years 1 and 2, there was no revenue, and hence no reduction in taxes for these expenses. The negative income, i.e., loss, in years 1 and 2 results in zero tax for those years, without accruing any tax benefit in subsequent years; then in later years the revenue was fully taxed. By taking depreciation in proportion to oil produced, we effectively shift the capital expense to later years in the accounts. The tax in the early years does not change (it is still zero), but the tax in the later years is reduced. That explains the increase in the simple, after-tax result. The second reason concerns the present value and IRR result, and is also an effect of timing differences. For these numbers, the basic issue is that a dollar tomorrow is not the same as a dollar today. A dollar today can be invested at interest, to give more than a dollar tomorrow; viewed in the other direction, it takes a little less than a dollar today to result in a dollar tomorrow, presuming interest is being paid. The amount that needs to be invested today in order to receive one dollar in the future depends on the length of time; the amount today is less for a 50 year investment than for a 5 year investment. As financial people say, "near money is dear money."

All financial models incorporate not just engineering estimates of costs, but also a number of assumptions about the policy and fiscal environments. Government policies with respect to tax treatment are only one policy area that can make a

Table 2.2 A slightly more complicated economic model of a one-well oil field

	Year 1	Year 2	Year 3	Year 4	Year 5	Year 6	Year 7	Year 8	Year 9	Year 10	Year 11	Year 12	Year 13	Year 14	Year 15	Totals
Revenue																
Oil (bbls per day)			250	219	191	167	147	128	112	98	86	75	66	58	50	
Price/bbl	90	90	90	90	90	90	90	90	90	90	90	90	90	90	90	
Sales revenue	0	0	21,60,000	59,13,000	51,84,000	45,09,000	39,69,000	34,56,000	30,24,000	26,46,000	23,22,000	20,25,000	17,82,000	15,66,000	13,50,000	3,99,06,000
Expenses																
Predrilling	5,00,000	10,00,000														
Drilling		50,00,000														
Completion			7,50,000		10,00,000											
Total capex	5,00,000	60,00,000	7,50,000	0	10,00,000	0	0	0	0	0	0	0	0	0	0	82,50,000
Royalty			3,24,000	8,86,950	7,77,600	6,76,350	5,95,350	5,18,400	4,53,600	3,96,900	3,48,300	3,03,750	2,67,300	2,34,900	2,02,500	59,85,900
Operating costs	5,00,000	5,00,000	7,16,000	10,91,300	10,18,400	9,50,900	8,96,900	8,45,600	8,02,400	7,64,600	7,32,200	7,02,500	6,78,200	6,56,600	6,35,000	1,14,90,600
Severence tax			2,16,000	5,91,300	5,81,400	4,50,900	3,96,900	3,45,600	3,02,400	2,64,600	2,32,200	2,02,500	1,78,200	1,56,600	1,35,000	39,90,600
Total opex	5,00,000	5,00,000	12,56,000	25,69,550	23,77,400	20,78,150	18,89,150	17,09,600	15,58,400	14,26,100	13,12,700	12,08,750	11,23,700	10,48,100	9,72,500	2,15,30,100
Total expenses	10,00,000	65,00,000	20,06,000	25,69,550	33,77,400	20,78,150	18,89,150	17,09,600	15,58,400	14,26,100	13,12,700	12,08,750	11,23,700	10,48,100	9,72,500	2,97,80,100
Results																
Operating results	−5,00,000	−5,00,000	9,04,000	33,43,450	28,06,600	24,30,850	20,79,850	17,46,400	14,65,600	12,19,900	10,09,300	8,16,250	6,58,300	5,17,900	3,77,500	1,83,75,900
Depreciation[a]	0	0	4,46,549	12,22,429	10,71,719	9,32,172	8,20,535	7,14,479	6,25,169	5,47,023	4,80,041	4,18,640	3,68,403	3,23,748	2,79,093	82,50,000
Profit before tax	−5,00,000	−5,00,000	4,57,451	21,21,021	17,34,881	14,98,678	12,59,315	10,31,921	8,40,431	6,72,877	5,29,259	3,97,610	2,89,897	1,94,152	98,407	1,01,25,900
Taxes (20% of profit)	0	0	91,490	4,24,204	3,46,976	2,99,736	2,51,863	2,06,384	1,68,086	1,34,575	1,05,852	79,522	57,979	38,830	19,681	22,25,180
Profit after tax	−5,00,000	−5,00,000	3,65,960	16,96,817	13,87,905	11,98,943	10,07,452	8,25,537	6,72,345	5,38,302	4,23,408	3,18,088	2,31,917	1,55,321	78,725	79,00,720
Metrics																
NPV @ 3%	−5,00,000	−4,85,437	3,44,953	15,52,828	12,33,136	10,34,218	8,43,726	6,71,237	5,30,755	4,12,563	3,15,055	2,29,794	1,62,662	1,05,766	52,047	65,03,302
IRR	70.5%															

[a]This is oversimplified, as the portion of the capital expense for installing the pump is included in the depreciation for the years before it is installed

difference. Similarly, changes in policy with respect to various operating procedures can change both cost estimates and other aspects of the model. As we have seen, such policy changes have the potential to result in major changes to the metrics of the financial model. It does not matter whether the policy is depreciation rules, tax rates or policies, regulations, subsidies, or whatever, the effect on the financial returns of a project can be significant. Often such policy changes are the result of governmental actions. And this introduces another dimension for risk: political risk. The existing government may "change the rules" by imposing new taxes, changing subsidies, changing regulations, and so forth. One major area of concern for oil and gas companies are the regulations for oil and gas operation: matters as how to dispose of water that is produced along with the oil and natural gas (there is always some), what must be done when a well is finally plugged and abandoned, what weight of trucks will be allowed on the roads, and a host of other issues. Additional political risk can be the changes of government (particularly in countries that do not have long histories of stable governments) and, in some areas of the world, armed conflict.

2.9 Additional Financial Considerations

The finances of our single-well oil prospect illustrate important metrics that oil companies use in assessing their overall projects and programs. Of course, nothing ever works out exactly as planned. Oil geologists are all familiar with "technical successes," where the exploration concept was essentially as predicted, the well found oil, but some detail meant that the oil could not be produced economically. While the geologist might consider the well a success, for the accountant this was a "dry hole."

For a company in the oil business, there will always be a number of prospects under evaluation. The business of the company is producing and selling oil, and so to stay in business it must keep finding replacements for the amount that it is currently producing. The process is continuous, with the projects that have the best financial prospects being the ones in which the company will invest its money.

So far our analysis has looked at finding and producing oil, but there is also the question of when does a well or project reach the end of its life. From our model, one can guess that it is when the project is no longer showing a profit. Because the capital costs are primarily at the beginning of the project, and because the amount of oil produced over the years declines, there will come a point at which the operating costs are no longer covered by the revenue being received. At this point, continuing to produce oil from the well creates losses for the company. Plugging and abandoning a well has some one-time costs; just as for the start of the project, these can be modeled and decisions taken in exactly the same way as for the start of the project. This course of action results in the best financial returns for the company. But unlike the new well, the costs of this operation have to be met from oil or gas that has already been produced.

Two important points need to be made about this end-of-life analysis. The first is that the decision will be made in context. For example, an offshore project will be analyzed on the basis of the entire production platform, not just of a single well. Offshore platforms are expensive to operate and maintain; the decision to decommission and abandon a platform obviously must apply to all the wells on that platform. Some onshore projects have similar infrastructure constraints; hence the operating costs of the Alaska pipeline become a factor when considering costs of continuing to produce oil from Alaska's North Slope.

The second end-of-life point is that just as projects can be bought and sold at the beginning of their lives, so too can this happen as they approach their end. Particularly with small, onshore wells in the USA, the company that started with the well is often not the company that finishes with the well. At any point during the production history of such a well, the future projected production, revenues and costs can be used to calculate project metrics from this time forward. Frequently a smaller operator will have lower overhead costs, and may therefore be able to profitably operate a well longer than a larger company. There is an environmental risk here; as production declines, the well may be successively sold to smaller companies that have lower operating costs, until eventually a well is sold to a company created just to buy this one well, produce it until its purchase price has been recovered, and then declare bankruptcy and walk away, without ever properly decommissioning the facility. Regulation can help, but is unlikely to completely resolve such problems.

2.10 Combining Prospects into Programs

The financial success of an oil or gas company depends on more than the financial success of a single prospect. Indeed, statistics show that US exploration wells are only approximately 50% successful (Petrostrategies 2012). This figure is substantially better than the exploration success ratio in the 1970s and 1980s, which was below 25% (Alfaro et al. 2007); the higher success rates are due primarily to advances in data processing capacity. Outside the USA the exploration success rate has generally been lower, although the fields may be larger The wisdom in the oil industry is that geologists have to have very thick skins because so many of their recommendations end up as dry holes.

But the higher success rate for the USA does not mean that more oil is found. In 2012 in the USA, the amount of oil discovered was 3 billion barrels, whereas in all the rest of the world it was 28 billion barrels, despite there being fewer wells drilled outside the USA (BP 2014). The conclusion is that, on average, the US discoveries were far smaller than the international ones. But remember that the goal of the company is not to drill wells that produce large volumes, but rather to drill wells that make money by producing oil, whether in small or large amounts. Thus the success of finding oil by a well is, for a company, only one part of the result; the total picture will be governed by the economic analysis of the costs to exploit the resource discovered and the projected revenues.

2.11 Finding Oil: A Risky Business

Drilling for oil or natural gas is sometimes considered a risky business. But over the past century it has not been so risky from a statistical viewpoint. The insurance industry and the gaming (gambling, bookmaking) industries are, in essence, the same business. Both make money by understanding the risk of a specific action and putting a price on a specific outcome. What makes them both profitable is that the risk profile is known. From actuarial tables or from bets which determine odds, the company can calculate the risk of having to make a payment and therefore determine a price for its service. This is risk. With risk, the outcome of a specific event may be unknown, but because the prediction can be quantified a price can be put on taking the risk. Traditional oil and gas exploration has been a risky undertaking.

Uncertainty is a different sort of thing. In the [in]famous words of Donald Rumsfeld, these are the "unknown unknowns" (Rumsfeld, 2002). Or as expressed by Laurence Peter of The Peter Principle, "some problems are so complex that you have to be highly intelligent and well informed just to be undecided about them".[4] There is no way to predict a chance of success because the parameters of the situation are not sufficiently understood to be able to make such a prediction. This distinction between risk and uncertainty and its implications in finance were described by Knight (1921) in what has become an economic classic.[5]

Given that oil exploration is a risky business, but not an uncertain one, the construction of a financially successful exploration program depends on applying appropriate risk adjustments to the financial model or models used. At a program level, "don't put all your eggs in one basket" is just as valid as for an investment portfolio. Portfolio management professionals have an entire library of methods that are used to design a well-balanced portfolio, but seldom do they discuss the case that a rather high percentage of the investments made will need to be written off entirely, which is the case for traditional oil and gas exploration. When an exploration well is a dry hole, the investment in it is for naught. The obvious conclusion is that the projects that are successful have to be very successful to compensate for the inevitable dry holes.

[4]Quoted in Mainelli and Harris (2011) p. 2.

[5]Uncertainty was a major topic at the time Knight was developing his ideas; his book predates Heisenberg's better known uncertainty principle in physics by 6 years.

2.12 Gambler's Ruin—The Risk of Failure

Mathematically one of the essential calculations for a company is how to avoid "gambler's ruin." This is the situation in which the person or company placing bets on known risks runs out of money before the known risks provide the projected return. To illustrate the point, in a frontier area if we estimate that the success rate of an oil exploration well is 15% it means that we have a failure risk of 85%. If we have $50 million to invest in the area and each well costs $7.0 million, then we can drill seven wells before running out of money. Our mathematical risk of running out of money without a success is thus

$$\text{Risk of complete failure} = (85\%)^7 = 0.85^7 = 0.32 = 32\%.$$

However, if we find another company that also has $50 million to invest in exploration, together we can afford to drill 14 wells and our risk of total failure is

$$\text{Risk of complete failure} = (85\%)^{14} = 0.85^{14} = 0.10 = 10\%.$$

To look at it another way, taking a partial interest in more wells will significantly lower the chance that we will go bankrupt before we find any oil. This explains why so many large exploration and development projects are shared between oil companies. The giant Kashagan oil field in Kazakhstan's portion of the Caspian Sea, for which cost estimates range from $46 billion to $116 billion (Demytrie 2012; Hargreaves 2012), has had at least eight large oil companies involved in the exploration and development at one time or another. As with the individual leasing arrangements discussed earlier with respect to ownership rights, the variety of arrangements, joint ventures, buy-ins, dry-hole contributions, etc., which companies use to balance their risk in such joint projects are infinite in their variety.[6]

Of course, sharing an exploration program with another company requires that the financial model developed in the last chapter will be different. Not only will it show only a portion of the costs because they will be split, but it will also only show the company's portion of the revenue. Furthermore, the starting point of our model is just that—for entire programs, the company will start by simply adding all the revenues and expenditures together, coming up with a combined financial analysis. Thus, one can speak of an entire program's IRR or NPV at 10% or whatever.

But to just combine more prospects together presumes that each has the same chance of success. In reality, we may have some prospects that we think have a 50% chance of success, others that have a 40% chance, and so forth. Combining these relies on the concept of expected value, developed by the seventeenth century French scientist and mathematician Blaise Pascal. Pascal developed the basic approach in order to better understand when to place wagers when gambling (Ore 1960). It is a simple concept: multiply the result by the chance of its happening in

[6]For details of the many ways an oil company can spread its risks see Quick and Buck (1983).

order to get the expected value. The expected value from our financial model can be combined with the expected value from other prospects to give an expected overall result.

The same concept can also be applied within the financial model for a single project. Thus, one might apply different chances to the price of oil, to the production rates, to the operating costs, and potentially to a host of other risks within the program. Using computers to test a variety of options quickly becomes a necessity.

2.13 Selection of Projects

What is important is that a company's management will select projects and programs that will provide the best financial returns to the shareholders. As the exploration budget is built up from individual prospects to the entire program the decisions will be made based on the combination of the potential of specific prospects as well as how well the specific prospect fits into the total program.

This bottom-up analysis skews a company towards making a "business-as-usual" set of investment decisions. So long as there are prospects and projects available for which the financial return is positive, this is where a company is likely to invest its money. As project competes against project for management selection, the people who work on specific projects will attempt to "sell" their best projects in this internal process. Indeed, careers frequently depend on successfully doing this. Each time a project comes up for review within the management decision process it will have gathered support from the people who have worked on it and believe it is worth the company's investment. In a large company this effect is replicated as various regional offices compete with one another for funding for their portfolios of projects. And in integrated oil companies, the exploration division must then compete against the refining division and the marketing division, because it is at the corporation level that there is a limit on the amount of capital available.

Management texts, MBA programs and management courses exist to teach ways of combining long-term strategic outlooks with the budgeting process. Most large oil companies are integrated, which is to say that they not only explore and produce oil and gas, they also ship it, refine it, and sell it. There is a story, believable but not verified, that a major oil company had a two-day board of directors meeting each October at which the budget, including the capital budget, for the coming year was decided. Each division of the company—the big requests for capital were from the Exploration and Production Division, the Refining Division and the Marketing Division—had worked long and hard on their presentations for the board, checking their figures and believing that their proposals would be of great benefit to the overall financial and strategic development of the company. So, of course, over the years they spent more and more time and effort on these presentations. Then one year an outside board member discovered that each division was hiring advertising agencies at the cost of millions of dollars for help in making their sales pitches.

Whether true or not, the tale is believable, because those working within a company are focused on their specific jobs. An exploration office of an oil company exists to explore for and find oil; when one has a hammer everything looks like a nail. Thus, the office will focus on finding oil and the associated planning of exploration and development of oil and gas discoveries; the employees in that office may go home to worry about alternative energy or climate change, but their daily job is to find and promote the best oil or gas prospects that they can. This description of decision-making explains why all industries, not just the oil industry, have a tendency to follow a "business as usual" path.

There is considerable debate both within and outside the oil industry as to how many good prospects are still available. As we will see in Chap. 4, we live on a finite earth, and at some point the supply of good prospects must come to an end. But the history of the industry is full of dire projections which have not come to pass. Today's view of the oil industry is shaped to a great extent by the period since 1945, and forgets that the world has been "running out of oil" before. Even in the period since 1945, the decade of the 1970s raised concerns that oil would never be plentiful again, only to have two decades of plenty from 1985 to 2005.[7] The high oil prices that characterized the industry from 2005 until 2014 again raised the question of what alternative sources for oil, and more broadly for energy, exist. At every turn these alternatives do seem to be expensive, although pursuit of them has resulted in at least a temporary oversupply of oil and collapse of prices. But the financial analyses show fewer and fewer good prospects that have the potential to replace oil production over the long term. So having rejected far-fetched alternative investments, many oil companies are turning to providing oil from "unconventional" sources, although as Berman (2015) has noted, "unconventional" in this context is basically a synonym for "expensive."

From initial idea to abandonment, oil wells and projects are undertaken based on the financial rewards to the company. At every point along the way, the decisions are made by making the best projections of revenue and expenses possible and then comparing opportunities. No company, or even country, can long survive if their decisions are consistently incorrect, which goes some way toward explaining why "business as usual" is the normal evolution of the economy. But this evolution is not always obvious—companies of any size are looking at multiple projects, thus making each company-wide forecast a complex combination of the individual pieces.

[7]Two interesting histories of the oil industry are Sampson (1975) and Yergin (1991).

Chapter 3
Some Basics of Petroleum Geology

The fossil fuels are the remains of plants and animals. Mother Nature recycles. Many living things become the food for something else. Plants become the food of herbivores, and herbivores become the food of carnivores. What does not get eaten by larger organisms gets eaten by smaller things: maggots and worms and such, which in turn are decomposed by bacteria. Throughout this cycle there is a flow of energy. Virtually all of the energy of life is captured by plants from sunlight and stored as chemical bonds between carbon and hydrogen and oxygen. An example of the basic reaction is

$$6CO_2 + 6H_2O + \text{sunlight energy} \rightarrow C_6H_{12}O_6 + 6O_2$$

in which carbon dioxide and water and sunlight create sugar and free oxygen. This is photosynthesis. When the sugar is needed for its energy, the reaction can be reversed by respiration or some other form of oxidation

$$C_6H_{12}O_6 + 6O_2 \rightarrow 6CO_2 + 6H_2O + \text{energy}$$

These chemical equations show only the basics. Detailed study provides many variants for different organisms, using different sugars and many other carbohydrates and resulting in different final products. For example, instead of respiration yeasts use the stored energy by converting sugar into alcohol and carbon dioxide by fermentation, a reaction that releases less energy per unit of carbohydrate than does respiration.[1] The important point is that the energy from sunlight is stored in the chemical bonds of the organic molecules and then released again. This generally happens at, or at least very near, the earth's surface. Most of the organic material formed by the energy of the sun is either directly or indirectly just recycled. But a small amount gets buried, and if everything goes just right this buried material can turn into a fossil fuel. Or, as it is sometimes called, fossil sunshine.

---

[1]This is probably why we respire (breathe) rather than ferment.

S.W. Carmalt, *The Economics of Oil*, SpringerBriefs in Energy,
DOI 10.1007/978-3-319-47819-7_3

A short digression: Some simple carbon chemistry

Carbon atoms can form many different types of chemical bonds, creating different materials. In a three-dimensional framework where each carbon atom is linked to four other carbons in a tetrahedral arrangement, the resulting material is diamond. When each carbon atom is linked to only three other carbon atoms, the result is that the carbon links form a two-dimensional plane, with each plane being able to slide easily between identical planes above and below, which is graphite.

Linked carbon to carbon in a chain, with all the links at the ends and edges being made with hydrogen atoms creates the simple alkane hydrocarbons as shown in Table 3.1.

The longer the chain, the higher the boiling point.

Most organic molecules in living creatures are considerably more complex, with tens and hundreds of carbons being linked to hydrogen, oxygen, nitrogen and smaller amounts of other elements in many different combinations and structures. Converting this material from its original structure into the structure of a fossil fuel requires geologic time along with moderate amounts of temperature and pressure. The liquid hydrocarbons, which we call oils, are an intermediate stage of this conversion from complex to simple.

The energy from fossil fuels comes from burning them. Chemically, it is combining the carbon and the hydrogen with oxygen of the atmosphere. Both the reactions, C+O_2 and H+O_2, release energy. But the chemical bonds of the original material must first be broken, so the total energy of burning is the sum of the energy released by the new products minus the energy required to

Table 3.1 Alkane (single chain) hydrocarbons

1-carbon	CH_4	Methane	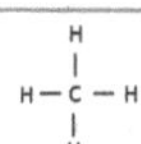
2-carbons	C_2H_6	Ethane	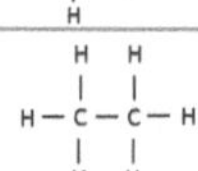
3-carbons	C_3H_8	Propane	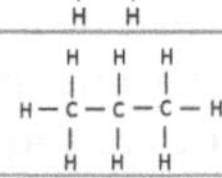
…			
8-carbons	C_8H_{18}	Octane	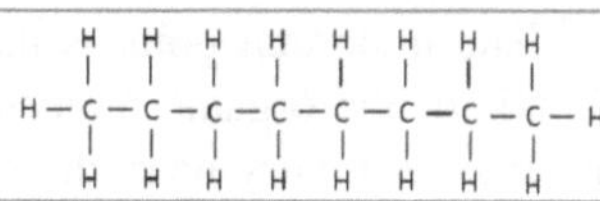
…			
10-carbons etc.	$C_{10}H_{22}$	Decane	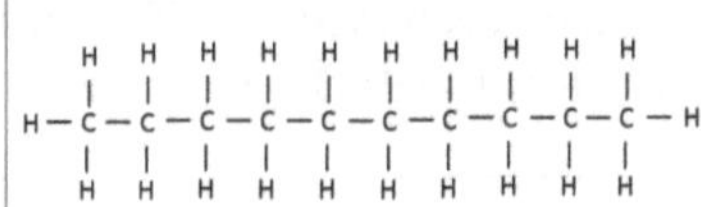

break up the original material. The $H+O_2$ reaction releases considerable energy, so the more hydrogen relative to carbon in the fuel, the more energy the fuel provides. The hydrogen to carbon ratio (H:C) thus gives a relative indication of how energetic a fossil fuel is. Natural gas, with a C:H ratio of 4 is a more energy-intensive fuel than octane with a ratio of 18:8 = 2.25. Coal, which is carbon rather than a hydrocarbon, has a ratio of 0. For a more complete and very readable discussion of the chemistry of burning fossil fuels see Courtney (2006).

Of the small amount of material that gets buried, the critical thing is that it be isolated from the oxygen of the atmosphere. If there is oxygen available, the material will generally just decompose and become part of the recycling biosphere. But for the small fraction that stays isolated from oxygen in a geologic area which is receiving sediments, the additional sediments bury the material deeper and deeper. This takes the carbon out of the cycle. The deeper one goes in the earth, the hotter it becomes, so in addition to being squeezed by the weight of the overlying deposits, the organic matter starts being cooked.

At this point, the formation of coal becomes a bit different from that of oil and natural gas. There are exceptions, but most coal is derived from land plants. The complex and robust structure of the cellulose of plants, in particular the cellulose which provides structural support, continues to provide strength during the early stages of burial. Furthermore, most coal comes from areas where the organic matter is not diluted very much with inorganic sediments (i.e., the plant remains are not mixed with mud). As the plant material gets buried the carbon atoms are packed closer and closer together. From the original organic matter, the deposit first becomes peat, then lignite, then bituminous coal and finally the hard anthracite coal, which is close to being pure carbon.

3.1 Conventional Oil and Natural Gas Formation

Oil and natural gas are mostly formed from one-celled marine plants and animals: algae, plankton, and diatoms. These organic materials have different organic structures from the cellulose that becomes coal. And, importantly, often they are buried together with considerable inorganic mud. Depending on the details of the original organic matter and the specific nonorganic sediments, this organic material will be reordered by the increasing temperature and pressure into different atomic arrangements. The resulting organic material is divided into two solid materials, kerogen and bitumen, distinguished by whether the material can be dissolved by organic solvents. Subjected to further burial, which increases the temperature and pressure, kerogen in particular continues its molecular rearrangement. In general, the molecular chains of carbons break apart, with the result that the material becomes more fluid. Thus as the temperature increases with the depth of burial the chains are first heavy oils, then shorter chain lighter oils, and finally gases. The last fluid is just the single carbon atom gas methane, after which increasing temperature results in only the carbon atom being left which is graphite, and no longer a fossil fuel.

Temperature acting over geologic time is important. The carbon in organic material is frequently found in long and complex arrangements of atoms which are broken apart ("cracked") with temperature. Some original organic matter becomes gas directly; other types of organic material initially become oil. The higher the temperature, the more cracking occurs. For the oily types of organic matter, the shorter the chains become. This gives rise to the concept of the "oil window," which is the depth to which the organic material must be buried to be converted into a liquid hydrocarbon. At a temperature of between 50 and 60 °C, which corresponds to a depth of between 750 and 1000 m, the right kinds of organic material will form liquid petroleum. The deeper the burial, and therefore the higher the temperature, the shorter the chains become until at about 150 °C, or between 4500 and 5000 m, the chains are gas. For the oil industry, the importance is that organic matter needs to be converted into a fossil fuel by a combination of pressure and temperature, operating over geologic time. These chemical reactions occur at depths that mean that pressures are also higher, due to the weight of the overlying sediments. But differences in pressure have relatively little impact on the result; it is mostly the effects of temperature and time that are important in the conversion. These factors have to be "just right"—the right type of organic matter, temperatures of between 50 and 150 °C and at least several million years for most oil. Natural gas is both an end product of the oil generating process and also is created throughout the temperature range, with some types of organic material creating more methane than other types.

In traditional oil exploration there are two additional requirements for creation of an oil field: migration and a trap. The organic material originally deposited is in very fine-grained rocks; if it were in coarser material there would be water circulating through it with oxygen. Once geologic time has slowly cooked the organic matter, more geologic time may cause some of the liquid to slowly seep into more porous rocks. Such a porous rock containing oil is called a "reservoir." But depending on the way the rock layers are oriented, the oil may simply seep through such reservoirs until it reaches the surface and is oxidized. This is what creates an oil seep, and oil seeps have been known from the beginning of recorded history. When the path of the seeping oil is blocked a trap is created. Not until the blockage is removed, for example, by erosion over geologic time, the oil collects there in the more porous and permeable rock waiting to be discovered. This explains the classic oil geologist's mantra: source, reservoir, trap. When looking at a prospect, that is what the geologist will be asking: is there a good source rock, organic rich that has been in the oil window to create oil?; is there a good reservoir rock into which this oil has migrated?; and is there a trap which makes this an economic accumulation worth drilling?

This migration of fluids into trapped accumulations is a geologic concentration of the fossil fuel resource. The resulting "pools" and "fields" of oil and natural gas can vary in size but they share the important feature that they contain oil and gas that has been concentrated in the location over geologically long times. Such fields are termed "conventional" oil or gas fields, and have been the source of essentially all of the oil and natural gas produced until very recently.

Not only has most oil production to date been from such conventional oil accumulations, also it has been preferentially from the largest. As we shall see in Chap. 4, these largest fields—called "giants" or "elephants"—account for the majority of both oil resources and oil production. And, after over 150 years of looking for them, most of the geologically likely places have been explored. So if oil and natural gas are to continue to be an important constituent of our energy supply, less conventional deposits will need to be discovered and exploited.

3.2 A Brief Excursion Through the "Unconventional" Alternative Sources of Oil

3.2.1 Conventional "Unconventional" Oil Accumulations

Often included with unconventional oil are fields in very deep water and in Arctic regions, especially in the Arctic Ocean. These are unconventional only because they involve very high costs; geologically, they are the same sorts of oil accumulations that have been exploited for over 150 years. As such, the projects are subject to the same sorts of risks as are all conventional oil projects, with the high costs putting huge amounts of capital investment at risk. The cost of ExxonMobil's single well in the Russian Arctic Ocean is reported as being $700 million.[2] If such wells are successful, they may cover the costs of a number of dry holes, but this is an exploration program that only companies with the very deepest pockets can undertake.

3.2.2 Tight Oil and Tight Gas

These accumulations are exploitation of oil or gas directly from the source rock. The oil is still in its original depositional environment. Because the organic matter was typically deposited in a muddy environment, the resulting rock has generally low permeability; i.e., the strata are "tight."

Ironically, these unconventional oil prospects have a lower risk of drilling dry holes, although it can happen. Rather, the risk is that the wells will not be as successful as planned, due either to technological problems or to a significant fall in

[2]Arkhipov et al. (2014).

the price of oil. Within the oil industry, there is major enthusiasm for tight oil as a future source of supply. The various technologies used to exploit tight oil have actually been used in the oil fields for decades, but have recently been combined to extract oil from strata previously considered as uneconomic prospects. For decades the oil industry has known of these shales in which oil is formed. Note, permeability is not the same as porosity; the permeability is the ease with which the fluid can flow through the rock, the porosity is simply the amount of space between the rock particles. The classical geologic history of oil and natural gas is, that over geologic eons, some of the oil from the shale has seeped into rocks of much higher permeability. Traditional oil fields have produced the deposits trapped in these high permeability rocks (reservoirs) by drilling a well down from the surface and through them, essentially vertically. Because of the high permeability, the oil or gas in the reservoir will flow some horizontal distance through the rock to arrive at the well.

But when the permeability is lower, various technologies may be used to enhance the permeability of the oil-bearing strata. Frequently, this is the technique called hydraulic fracturing, or simply fracking. It has been used for decades in vertical wells to enhance the permeability of a natural reservoir into which oil has migrated. When permeability is especially low, the reservoir is "tight," because the porosity is so low that oil does not flow readily. When combined with techniques that allow the well to be drilled along the geologic strata ("directional drilling") a greater length of the well can be in contact with the oil-bearing rock. Greater length with low porosity can thus combine so that the well produces sufficient oil (or gas) to be economic, even when each unit length does not generate much oil. Sometimes this may be a rock into which oil has migrated, but more recently this tight oil extraction method[3] has been used to directly exploit the oil that is still in the source rock, as shown in Fig. 3.1.

After the well has been drilled, the impermeable shale has its permeability increased by fracking, which injects water into the strata under very high pressures to fracture it, thus artificially increasing the permeability so that oil or gas can flow to the well. The fracking technologies have actually been used in the oil industry for decades. As with all technologies, they have been improved with time. The water used has sand or ceramic grains added to it to keep the new cracks open once the pressure is released, and has some additional chemicals added, designed to give the water the precise physical characteristics needed to be most effective.[4] When oil prices were very high, between about 2007 and 2014, the technique was deployed much more extensively, and much controversy resulted.

---

[3]Technically, "tight" simply means that the strata has low permeability, so there are situations where the oil has seeped from the shale into another impermeable layer which is not shale. Another situation is for the shale layers which contain the organic source material to be interbedded with thin, impermeable siltstone or limestone layers which have low permeability. All of these are tight oil deposits.

[4]Fire fighters also do this, adding small amounts of a polymer to water so that it will have less friction going through the fire hoses and hence reach further into the fire. In effect, they make the water more slippery, see Gleick (1988).

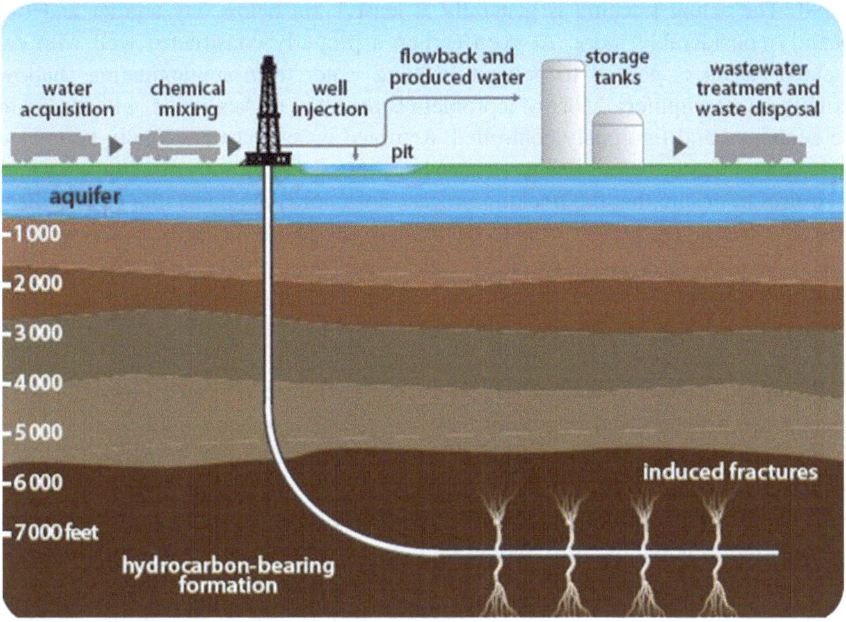

Fig. 3.1 Horizontal drilling and fracking (*Source* Welch 2014)

For oil economics there are several issues: the first is that the fracking process is expensive in both monetary and energy terms. As one might expect, the amount of pressure, and hence energy, required to frack the rock is considerable, and increases with depth. This increase with depth is one of the reasons that some shale basins are being reevaluated with respect to their present economic viability. A second problem is that the amount of water needed is considerable. Amounts will vary, but the figure often cited is 15 million liters per well (McGrath 2013).[5] Since a tight oil or tight gas area will require hundreds or even thousands of wells, the quantities of water needed add up, causing a problem in some of the tight oil and tight gas provinces that have been identified. The proposed large tight gas resources of both Argentina and China, considered among the largest in the world, are in arid or semi-arid areas.

Much of the water used in fracking returns to the surface after the pressure of the fracking operation is removed. This water can be recovered and recycled which cuts down on the amounts that are actually consumed. Transportation from one well to the next is often accomplished with temporary pipelines to reduce the wear and tear on the local road infrastructure. But the water that returns to the surface from the fracking operation is often contaminated by the high-pressure contact with rocks at

[5]An Olympic standard swimming pool (50 × 25 × 2 m) holds 2.5 million liters; so visualize 6 Olympic swimming pools worth of water needed to frack a well.

depth. The actual fracking is generally at least 1 km below any aquifer and frequently considerably more. As seen earlier, a properly constructed well with cemented casing is needed to prevent fracking water from contaminating shallow drinking water aquifers. More of a problem is potential surface spills, which need to be contained and rigorously controlled. Returned water must eventually be treated as chemical waste[6] before being released back into the environment. All the various measures to ensure that the fracking process does not contaminate the environment cost money, and these costs need to be added to the economic analysis. The general effect, of course, is that such increased costs lower profits, thus changing the economic profile. Much of the political debate about fracking is, fundamentally, about the costs that will be required to contain and treat the water that returns to the surface.

Another problem with tight oil deposits is that the amount of oil that can be produced from each well is usually more limited than in conventional oil and gas wells. This is due to the low permeability of the formation. The horizontal drilling technique, which exposes the well to a much greater volume of oil-saturated rock, and the fracking technology, which provides some high-permeability channels between the well and the surrounding rock can only partially solve these problems.

Yet another issue is the amount of the oil in the rock that will flow to a well, however slowly. All oil and gas wells leave a lot of oil and gas in the geologic strata, but how much depends on the size, shape, and chemical composition of the rock particles. What happens is that the oil or gas molecules "stick" to the rock particles, and this 'stickiness' depends on both chemistry and physics. Sometimes the molecules that stick to rock particles actually shut off pathways to the well, decreasing the permeability. As one might guess, this is more of an issue in the tight reservoirs where the spaces between rock particles are much smaller than in more conventional reservoirs. Production rates typically decline much more rapidly in tight reservoirs than they do in more porous, conventional reservoirs.

The result is that tight oil or gas development means both that more wells are needed for the volume of rock, which costs more, and the amounts produced will fall off rapidly, which results in less revenue. Overall, a lower percentage of the original oil or gas will ever arrive at the well.

Putting this all together, it is clear that tight oil and gas wells are going to be a considerably more expensive means of extracting hydrocarbons than traditional wells. In a market economy, if it makes money someone will do it. When oil was trading above $80 per barrel such projects did make money in the US and the result was that US production increased for the first time in decades. When oil prices sank to below $50 per barrel this activity effectively ceased. The wells already drilled continue to produce, however. From the company's viewpoint, it is simply a matter of whether the well will be profitable as outlined in Chap. 2. From a broader

[6]Depending on the shale, the contaminants may include chemicals that render the water carcinogenic if subsequently treated to be used as drinking water, or in other cases may be made radioactive (Warner et al. 2013).

Table 3.2 Price return estimates for a typical Bakken oil well (based on initial production and decline rates given by Hughes (2014))

Oil price ($/bbl)	Before tax profit ($million) after 15 years	IRR	Pay out (year)
100	2.8	11%	5
90	1.5	6%	6
80	0.2	1%	9
70	−1.0	Negative	>15

economic perspective the question is whether such deposits have the potential to replace the oil and gas produced from the more traditional sources that are now beginning to decline.

With respect to the first question, whether an individual well will be profitable, the answer is found by using the same financial model as was used in Chap. 2. While individual wells will vary in details, the information available indicates that most wells are, indeed, profitable. For Bakken tight oil wells Hughes (2014) gives typical starting production values and decline rates. Data available only extend about 5 years, but projecting to a 15-year lifespan of a well the results are shown in Table 3.2.

It should also be remembered that the costs are mostly incurred in drilling the well, with the revenue stretching out for a number of years afterwards. When oil prices fall after production begins and the amounts being produced are falling as well, the project may change from being profitable to being unprofitable. Thus, the oil price collapse of 2014–2015 has made many existing wells unprofitable, which has, of course, discouraged drilling of new wells in tight reservoirs. But because the bulk of the spending has already been done, the company is likely to continue to produce as much oil as possible from the wells already drilled.

A similar analysis can be done for tight gas wells. There is an additional complication here in that there is a considerable range in the initial production rates from which the declining production is calculated. Within the industry a high initial production rate is frequently referred to as "hitting a sweet spot." The price for gas has fluctuated around $4 per MMBtu since 2009, more often dipping below the $4 level than rising above it.[7] Initial gas production rates vary widely, but many wells in the Marcellus formation in Pennsylvania have initial production rates of between 4000 and 8000 Mcf per day, although some wells have considerably higher rates. Using decline rates given by Mearns (2013), 15-year break-even points for single wells can be calculated as shown in Table 3.3.

Tight oil and gas developments on a wide scale are a recent phenomenom, so 'typical' declines beyond about 5 years are still estimates.

The second question about tight oil and gas is whether it will be able to compensate for the declines that are showing up in more conventional oil fields. The outlook for this is not so good. In a conventional oil field, production begins and

[7]A number of financial services provide price quotations, for example http://www.nasdaq.com/markets/natural-gas.aspx. Consulted 2015-01-27 for the figures quoted.

Table 3.3 Break-even initial flow rates for Marcellus gas (calculated from data in Mearns (2013))

Initial production rate (Mcf per day)	Gas price required for 15-year break-even
4000	$4.20
5000	$3.30
6000	$3.10
7000	$2.40
8000	$2.10
9000	$1.90
10,000	$1.70

frequently increases for a number of years because more wells get drilled in the field and individual well decline rates are low. But these conventional wells show declining production over time, and the need is to replace these declines. Because tight oil wells show rapid declining production rates almost immediately, the only way to outpace the declines from conventional oil fields is to drill new wells faster and faster. As soon as new wells stop being drilled, the production from the entire tight area will start to decrease and thus will no longer offset declines from conventional fields. The faster and faster drilling of tight oil (and gas) wells is referred to as 'factory drilling', where the operation is designed to drill wells in rapid succession so that each new well can replace the declines from the wells just completed (Forbes and Wilcyznski 2010). Numerous critics observe that the statistics showing increasing production from tight oil plays is the result of more wells being drilled, and it is obvious that this cannot last for too long. Just keeping production constant is subject to the Red Queen effect,[8] with increasing production requiring ever increasingly fast drilling of new wells. The fact that there must be a limit to the extent of the oil-rich strata has led to cautions on the ability of tight plays to offset declines in more conventional fields (see, for example, Likvern 2012, and Hughes 2014). Furthermore, the extent to which companies expand their activities so as to actually increase production using borrowed money raises the spectre of a financial bubble (Berman 2010).

The tight oil and gas resources exist in the ground. The question is whether they can be economically exploited. As is implicit in the way that tight wells are drilled and fracked, a great deal more energy is used to make a well in a tight oil or gas play than in a conventional oil or gas field. Chapter 6 will discuss such energy balances in greater detail. But it is obvious that the energy return, more precisely the energy return on energy invested and generally shortened to EROI (Cleveland et al. 1984),[9] is lower for tight oil exploitation than for conventional exploitation.

---

[8]The Red Queen Effect is named for the Red Queen in Lewis Carroll's *Through The Looking Glass* (1871). The Red Queen says "Now, here, you see, it takes all the running you can do to keep in the same place. If you want to get somewhere else, you must run at least twice as fast as that!".

[9]According to Murphy and Hall (2010) the Cleveland et al. article in 1984 seems to be the first use of the specific EROI acronym; the concept had been articulated earlier, e.g., by Hall (1972) and Hall et al. (1979a, b).

Estimates of EROI for tight oil are given by Hall and Klitgaard (2012) as being between 5:1 and 12:1; this compares to figures of over 25:1 for almost all conventional oil and natural gas fields. The relatively shallow (i.e., low drilling costs), and permeable oil fields of Saudi Arabia are estimated to have original EROI values of approaching 100:1. As with all resources, the best deposits are the first ones exploited (Bardi 2014), so the EROI of oil and gas discoveries decreases over time. One question about the tight oil technology is whether it simply represents lower EROI values that are expected as easier oil deposits become more scarce or whether tight oil represents a discontinuity in oil occurrence which also requires a much lower EROI technology.

Because crude oil is the output of both conventional and tight oil plays, they can be compared without additional calculations or computations. The details of drilling and completing the individual wells may have technical differences, but there is nothing fundamentally different in the way the economic evaluations are made. One of the major factors differentiating the two types of prospects is the difference in the risk profiles. For tight oil finding the oil-rich strata is fairly predictable, with the risks being in the technology and the decline rate; for conventional oil fields there is more risk about whether the well will penetrate a sufficiently permeable, oil-bearing geologic strata and find oil, with the extraction technology and decline rates better known from experience.

3.2.3 *Other Unconventional Oil*

Two even more unconventional hydrocarbons are potentially available, and, to a somewhat limited degree, are being used, as oil sources. The first are the very viscous tars that are found in abundance in Venezuela, Canada, Russia, Kazakhstan, and Malagasy. These tar sands[10] are oil fields that have such heavy, dense oils that they cannot be produced by normal oilfield methods. Either the deposit must be mined and processed, with technologies similar to many other mining operations, or an in-situ processing can be done to allow extraction. Either way, the goal is to break the long hydrocarbon chains that characterize tar into the shorter chains that are crude oil. After these steps, the result is a fluid crude oil which is transported to traditional oil processing facilities.

Whatever methods are used, the costs of extracting crude oil from tar sands are significantly higher than for conventional oil or even tight oil; this is true whether one is speaking of the financial cost or of the energy cost. The reason is that in addition to whatever energy inputs are needed for the initial extraction, there is an additional energy input needed both directly and indirectly to break down the tar into something that is similar to a conventional crude oil.

[10]Canada and the oil companies prefer the term "oil sands".

The attraction of the tar sands is that the calculated reserves are so large[11] that only the constraint of low EROI limits their role as an available oil source. As a result, each of the major international oil companies (ExxonMobil, Royal Dutch Shell, Chevron and BP) have joint venture operations in the Canadian tar sands. In part this may represent a willingness to accept somewhat lower financial returns on their capital investment in order to acquire the experience needed to be able to continue with successful investments with this resource.

Somewhat similar from an economic perspective are oil shales. But just as tar sands represent a step down in EROI from conventional oil (i.e., they require more energy input per unit output), so oil shales represent an additional step down in EROI from tar sands. The reason is that while an oil shale can produce oil, it is a qualitatively different resource from both natural crude oil and from tar. Oil shale is shale rock that includes significant organic material which only has the potential to be made into oil. Oil shale is not to be confused with shale oil—the two are different. Oil shale is shale rock in which the organic material has not yet become oil, whereas shale oil is oil found in shale rocks and is produced from these impermeable ("tight") formations as already discussed. Shale oil is a form of tight oil, extracted from the deposit by pipes; oil shale is rock which needs to have the organic matter converted into oil. Very simply, oil shale is a potential source rock which needs a bit more "cooking" to separate the oil from the kerogen.

Not only is considerable energy needed to finish the process of converting the kerogen into oil, but the heat applied to the oil shale in order to do this has the unfortunate side-effect of changing the mineral structures in the rock, with the result that the residue material occupies more volume than the original material. While companies periodically evaluate the Green River Oil Shale,[12] none has yet found it to be an economic competitor to other oil sources, even when compared to tar sands. Given that oil shale is even less economic to exploit than is tar sand, and given that there are very large quantities of tar sand, oil shale is unlikely to become competitive in the near future.

[11]Canada and Venezuela have at least 300 billion barrels of reserves primarily because of their large tar sand deposits; this compares to 55 billion barrels for the USA (BP Statistics 2014). These large reserves, available at known costs, are the economic equivalent of a chemically buffered solution. Economically, the EROI for oil shouldn't go below the EROI for tar sands—if it does, more tar sand oil will be produced holding the EROI to that value. Because the tar sand reserves are there and the entire technology is known, bringing more oil into production is only a timing question of opening more mines and constructing other needed infrastructure.

[12]The Green River oil shale, located in the area where Colorado, Utah and Wyoming meet, is one of the best understood and largest of oil shale deposits. It has been known for a long time, because when early trappers used it to surround a campfire, the heat from the fire could finish the conversion of kerogen into oil and the rock would "burn". While the calculation of potential hydrocarbon resource is very large, no matter what the price of oil, oil shale always requires a price that is a bit higher to make economic sense.

3.3 Non-oil Energy Options

The non-oil fossil fuels, coal and natural gas, tend to have somewhat lower energy costs than does oil. But while the energy per dollar may be greater, the real cost of using any particular energy source must also include a number of additional factors. For example, coal generally has a lower unit energy cost than oil, but there are additional costs associated with pollution control and conversion of coal into other forms of energy. Do not expect to see steam locomotives returning to the rail system. As we will see in later sections, natural gas has significant costs of moving the resource to markets. Hydroelectric power is also inexpensive on a per unit of energy basis, but large quantities of electricity cannot easily be stored. What follows is a brief listing of the major sources of non-oil energy.

3.3.1 Natural Gas

While a fossil fuel, natural gas can be used as a replacement for oil in many applications. Natural gas engines for cars and trucks are very similar to existing gasoline and diesel engines; sometimes it is even possible to convert an existing engine. Not only is natural gas less expensive on a per-unit energy basis, but also it produces less carbon dioxide per unit of energy. Many oil wells produce natural gas as well as oil, and through the history of the oil industry untold amounts of natural gas simply have been flared (burned) as an uneconomic byproduct of oil production. Some types of organic matter produce much more gas than oil, with the result that there is no shortage of natural gas. For many years, one of the reasons for a 'dry hole' when drilling for oil was that gas was found instead. Oil company files have information about many wells that found gas instead of oil, and which were therefore plugged and abandoned.

The disadvantage of natural gas is that it is a gas. It is much easier to put a liquid (oil) into some sort of tank which is then transported to the user than it is to do this with gas. But because of the similarity of oil and natural gas, and the fact that many oil wells also produce gas, oil companies are almost always natural gas companies as well. Development of prospects is entirely similar to developing oil prospects, and project evaluations are the same. Natural gas will be a part of the future of today's oil companies, as Chaps. 7 and 9 discuss in greater detail.

3.3.2 Coal

Coal was the fossil fuel first exploited. Initially lumps of coal were picked up where coal strata are exposed at the surface. Coal's geologic origin has some similarities to oil and natural gas, but is distinct from them. Unlike oil and natural gas, which have their origins in microscopic bits of organic matter which are primarily derived from single-celled organisms, coal is derived from much more complex woody plants. While some oil and natural gas source beds are from lakes, most are from a marine environment; by contrast, the woody coal organic source material is primarily from swamps and bogs, which are terrestrial environments. One can follow a complete gradation of coaly deposits ranging from peat, through lignite, bituminous and eventually anthracite coal, with the last being formed at temperatures and pressures greater than those leading to natural gas. Because of its more complex organic source, coal tends to have many more trace impurities included in it than do either oil or natural gas; this can make coal extremely polluting when burned for energy. Furthermore, while some of the most modern coal uses are fairly efficient, the traditional steam engine is not; the most modern efficient coal burning requires large capital investments and mostly run on an almost continuous basis to achieve their efficiencies.

Because the initial deposits of coal were found at the surface, it was relatively easy to dig it out. And coal is easy to transport. The digging progressed to become mining operations. The original steam engines, and hence in some ways the entire industrial revolution, were developed in order to pump water from coal mines.

3.3.3 *Nuclear*

Nuclear energy has the advantage that it does not produce carbon dioxide. But it does produce very long-lived radioactive waste. Furthermore, when something goes wrong, a significant surrounding area may be subject to radioactive contamination. In fact, nuclear reactors have a better safety record than most fossil fuel systems, but public perception is the opposite.

3.3.4 *Hydroelectric*

As already mentioned, hydroelectric power is inexpensive energy, but the electricity generated must be used as it is generated because storage of electricity is expensive. This also means that most applications need to be directly connected by wires to the generating plant, which limits major mobile energy uses, for example in trucking and farming. A further limitation is that many of the best hydroelectric sites, i.e., good locations for dams, have already been used.

3.3.5 *"Renewables"*

A host of technologies are rapidly evolving to harness solar energy directly (e.g., photovoltaics) or indirectly (e.g., wind). The costs of these have been rapidly diminishing, but as yet they are not producing quantities that are sufficient to replace very much oil. This may be changing, but any realistic projection of displacement still must be measured in decades rather than years.

Oil companies operate within the context of society as a whole, and oil must compete economically with any other energy source. In general terms, comparisons can be made of the EROI of alternative fuels, although the details make such calculations frustratingly problematic. Energy from nonfossil fuel sources presently provide less than 20% of what the global economy needs (BP Statistics 2014). Of the nonfossil-fuel energy, nuclear and hydropower each contribute more than 5%, leaving all of the other renewables—wind, solar, biomass, and everything else one can think of with the final less than 10%. Subsequent chapters will examine the economic issues of these alternatives for the oil industry in greater detail.

Chapter 4
Peak Oil

The history of the oil industry is one of gluts followed by concerns over future supply. Major oil discoveries have tended to flood the market with supplies far in excess of demand. But the resulting low prices increase demand, and as the oil stops gushing out of the wells there is concern about shortages. By the end of the nineteenth century the Standard Oil Trust was relatively successful in smoothing out these cycles. But with the breakup of trusts in the early twentieth century, the problem reappeared. With the increasing importance of oil the role of price maintenance shifted to some extent to governments. But when output from the new discoveries began to decline, as it always did, the concern turned to what would happen when the oil ran out. In 1919 the Chief Geologist of the US Geological Survey predicted that peak oil production would occur within 3 years (White 1919).[1] These ups and downs of the oil industry make for exciting reading, complete with colorful personalities. Anthony Sampson's *The Seven Sisters* (1972) and Yergin's *The Prize* (1992) are good examples. In the oil fields the reality was that discoveries outpaced the rapidly increasing demand, although the path wasn't even.

4.1 Hubbert's Predictions of Oil Supply

As new oil province followed new oil province allaying fears of shortages, and just as the vast oil resources of Saudi Arabia were becoming evident, the geologist M. King Hubbert (1956) predicted that this would all end—and soon. In a presentation to an industry conference, Hubbert made the bold prediction that the onshore production in the USA (excluding Alaska) would peak in about 1970. That was only 14 years away, the short time being a consequence of the exponential growth in consumption that was still occurring. That US onshore production did peak in 1970 has given Hubbert the honor of inventing the "peak oil" concept.

[1]As quoted in Ahlbrandt (2012).

S.W. Carmalt, *The Economics of Oil*, SpringerBriefs in Energy,
DOI 10.1007/978-3-319-47819-7_4

Hubbert's major point in 1956 was that exponential growth means that the exhaustion of a finite resource will come with surprising speed. If production grows at 7.9% per year, the absolute amount produced doubles in less than 9 years. The title of Hubbert's 1956 paper was *Nuclear Energy and the Fossil Fuels*, and his major point was that to ensure sufficient energy it was none too soon to begin an all-out transition to uranium and nuclear energy.

The concept of a peak comes from Hubbert's theoretical model of any finite resource exploitation, as shown in Fig. 4.1.

The popularity of the term "peak oil" is more recent, probably dating to press reports about the founding of an Association for the Study of Peak Oil in 2002.[2]

Hubbert had long been interested in the finite nature of earth's resources,[3] and his first paper on the finite nature of oil is from 1938 (Hubbert 1938). In addition to his 1956 prediction of the US peak oil production in about 1970, he predicted that world peak oil production would occur in about 2000. Hubbert's 1956 publication predicted the year of peak production; but to make this prediction he needed estimates for the total amount of oil that would be produced. In his 1956 paper he used estimates for this Total Recoverable Resource that had been made by others[4]. Hubbert showed how producing that quantity of oil given the production history to date necessitated reaching a peak in production sooner than people might expect.

Hubbert continued to work on the problem of oil supply, and his subsequent papers (Hubbert 1962, 1967, 1969, 1974, 1982, amongst others) used a variety of techniques to estimate both the ultimately recoverable resources and the approximate date for the peak in production. Perhaps the most important insight was his realization that oil must be discovered before it is produced and hence the information from discoveries can also point to the total amount of oil before the production information provides the same result.

Dr. Hubbert was a brilliant scientist. A recent biography (Inman, 2016) chronicles much of the development of his thinking about oil and other natural resources and details the dissention that Hubbert's ideas caused within the industry and within the US Government. One of Hubbert's greatest strengths was to simply plot the data and present it with what, to him, were obvious conclusions—and then not back down in the face of political pressure. This did not make him popular in many circles, but he had the good luck to have supportive employers for most of his career. And, as he would point out, he was only collecting data and presenting it—if anyone could show that his data was incorrect or that his reasoning was faulty he would retract his opinion. This seldom happened, although he was as critical of his own work as he was of the work of others.

[2]See Aleklett (2012) p. 10 for a history of the term.

[3]Kuykendall (2005) has compiled a bibliography of Hubbert's publications. See http://www.hubbertpeak.com/hubbert/bibliography.htm. Accessed 2014-09-02.

[4]Notably those of Lewis Weeks, chief geologist of Standard Oil of New Jersey (now ExxonMobil).

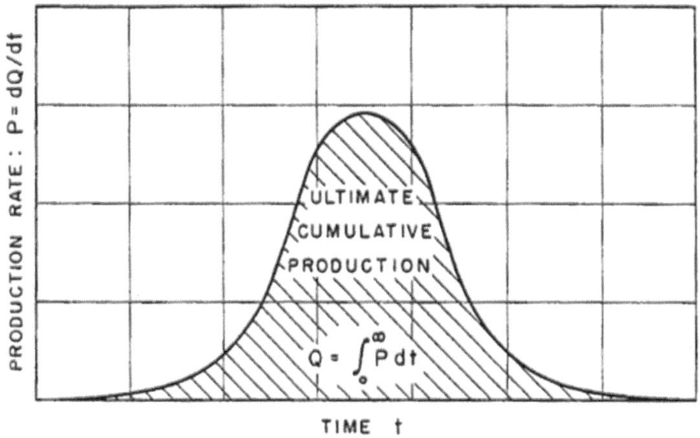

Fig. 4.1 Production of a finite resource (Hubbert 1956)

Hubbert's predictive insight, as he put it, was that when the last drop of oil is produced, the amount produced will be equal to the amount discovered (Hubbert 1962). Using this insight, Hubbert was able to make his own projections for the total amount of oil available from published data on discoveries, rather than relying on estimates from others as he had done in 1956. He then proceeded, as before, to use this total resource number to predict when the peak production rate would occur. Although Hubbert continued to explore the topic, the essence of his theory was in place by 1962. This did not please the political and business powers within the oil industry, who continued to proclaim that there would always be enough oil; industry spokesmen argued that if oil became scarce the price would go up which would result in more drilling and more oil being discovered. So little was done to address the concerns that Hubbert raised.

A younger colleague of Hubbert's, Ken Deffeyes, dug through Hubbert's mathematics and published Hubbert's technique in two books (Deffeyes 2001, 2005). In addition, Deffeyes validates some of Hubbert's work by examining the oil discovered in Kansas,[5] which is important because it ties Hubbert's methodology to statistical sampling. Using Hubbert's logistic assumption with known production data, Deffeyes was able to make new global projections: ultimate recovery of 2000×10^9 barrels with a peak production date of November 24, 2005.[6]

[5]Kansas has been very thoroughly explored for oil. This allows statistical forecasting methods to be tested against known results.

[6]Deffeyes states that the uncertainty in the analysis is ± 1 month, allowing him to choose US Thanksgiving Day of 2005 as the actual date of peak oil (Deffeyes 2005, p. 43).

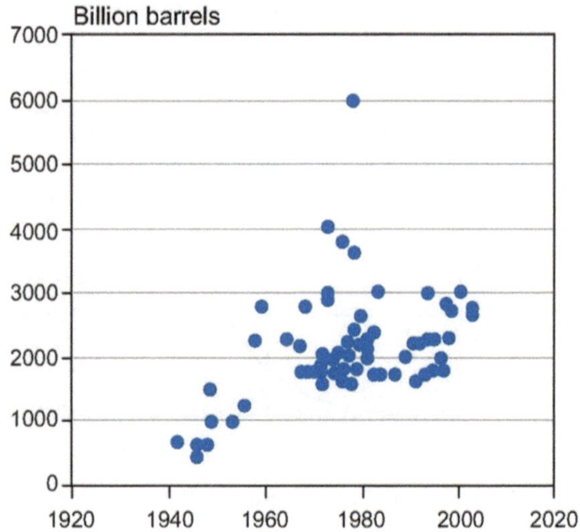

Fig. 4.2 Estimates of ultimately recoverable oil (from Sims et al. 2007)

It is important to note that Hubbert and Deffeyes are using statistical methods to reach their results. Statistical methods are used frequently in science. Deffeyes noted a saying amongst statisticians "you don't need to eat the entire ox to know that the meat is tough".

Many in the oil industry found, and continue to find, Hubbert's forecast unsettling. The intellectual excitement for the exploration part of the oil business was, and continues to be, figuring out new places to drill for oil. And, while exploration or "wildcat" wells have a low success rate, the successes can be spectacular when they come. Most people in the industry will acknowledge that, given that the earth is finite, a peak must exist. But they question its timing, the total amount of oil that will be produced, the shape of the production curve, and other aspects. Very simply, if the total amount to find is larger, the peak date can be farther in the future. So many, perhaps most, industry analysts have argued that the ultimate recovery figure is larger than Hubbert's or Deffeyes' calculations. Many of the arguments date back to the 1960s in the wake of Hubbert's paper, and are still being used today. The range of a number of oil estimates is shown in Fig. 4.2.

The 1970s saw a more widespread acceptance of Hubbert's thesis, although many in the oil industry remained skeptical.[7] Numerous other forecasting techniques were used, for example calculating the amount of oil drilled per foot of exploration wells drilled, or doing volumetric calculations of entire basins. In addition to the debate surrounding any new insight, Hubbert's papers created a stir

[7]A conference in 1974 (for proceedings see Haun (ed) 1975) was held at Stanford University on the subject of 'Methods of Estimating the Volume of Undiscovered Oil and Gas Resources'. Hubbert was conspicuous by his absence.

because of their timing. NASA's first photographs of "Spaceship Earth," the Club of Rome's *Limits to Growth* (Meadows et al. 1972) report, and gasoline lines resulting from the Arab embargo all combined to create a sense that Hubbert's theory had merit. Then, in the mid-1980s when the price of oil fell, there was more than ample supply with plenty of oil available, interest waned in projecting when this finite resource would run out.

A short digression: Resources, reserves, and rates of production Discussions of peak oil, and of the policy question "are our oil supplies adequate?", often come apart over the differences and relations among the terms *resources*, *reserves*, and *rates of production*. One might call these the "three Rs" of oil policy.

The oil and gas industry has adopted a common vocabulary to describe the various types of oil and gas which the various companies, based on their specific surveys, think is in the ground. Because oil and gas is found in rocks thousands of feet below the land surface, calculations of how much is there are subject to many types of uncertainty. In Table 4.1, which is the industry standard, this type of technical uncertainty is expressed on the horizontal axis.

Table 4.1 Oil and gas industry definitions for reserves and resources (SPE 2011)

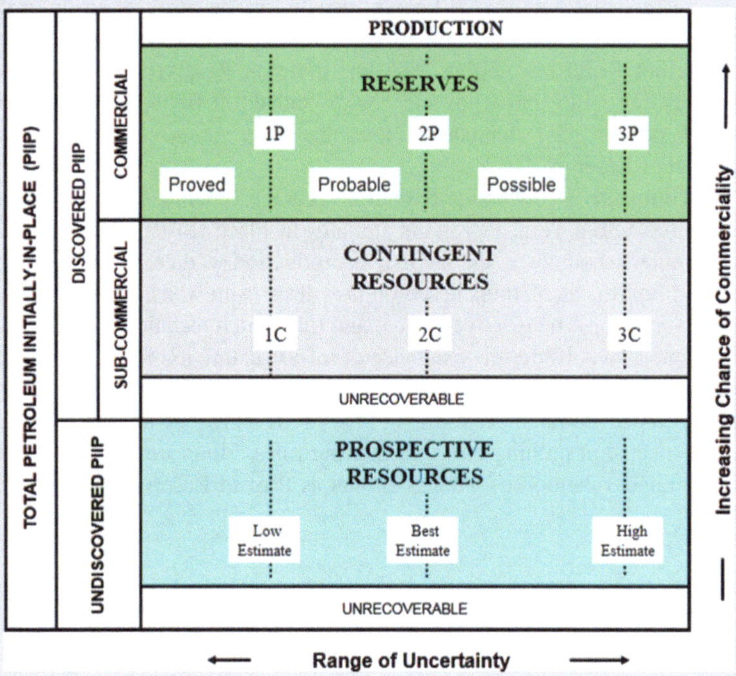

We have seen that the amount that may be economic to produce will depend on many factors, any of which may change. This is incorporated in the vertical axis together other uncertainties. The total oil or gas that was originally in the rocks is the area of the table. The amount produced so far is known, so that is the solid white bar across the top. After that are the important definitions.

Reserves are oil or gas that has been discovered and evaluated as being economic to produce ("commercial") based on wells having been drilled. In conjunction with the detailed seismic surveys that have been done around the well, the likelihood of a well producing more oil can be calculated. The technical uncertainty in the evaluation increases with the distance from the tested well and with many other specific factors. Taken together, the amount of oil for this known accumulation can be divided into proved (>90% certainty), probable (between 50 and 90% certainty), and possible (less than 50% certainty)[8] reserves.

Resources is a more encompassing term. The SPE guidelines consider resources as being distinct from reserves, as shown in Fig. 4.3. Unfortunately, the word is frequently used collectively to include reserves as well. The agreed SPE definition is that resources do not include reserves. They are divided into two categories: Contingent and Prospective. The Contingent Resources are reasonably well-known, usually as the result of drilling; it is just that at the current price of oil or gas they are not economic and are therefore not worth developing. The line dividing Reserves and Contingent Resources can easily move up and down—indeed, it frequently does as the price of oil (or gas) fluctuates. Note that there is no such thing as a "Contingent Reserve".

The Ultimately Recoverable Resources (URR) sometimes discussed is the same as the SPE's Total Petroleum Initially In Place (PIIP), and Remaining Recoverable Resources is the URR less production to date.

In addition to calculations based on flow tests from wells, interpretation of seismic surveys of the discovery area, and other such technical criteria, these evaluations also consider the existence of infrastructure to bring the oil or gas to market and the legal ownership status of the oil or gas. The SEC[9] allows publicly traded companies to show Proved Reserves as an asset on their balance sheets but nothing else. Accounting rules[10] thus are important in the definition and valuation of what is shown as Proved Reserves.

[8]Referred to as 1P, 2P and 3P reserves, with 1P being the proved reserves.

[9]Securities and Exchange Commission, which regulates securities trading in the USA.

[10]See Financial Standards Accounting Board ASU 2010-03 (FASB 2010).

Rate of production is what it purports to be, the rate at which oil or gas is produced. But dividing reserves by the rate of production does not usually give the time remaining before the field stops producing. The rate of production declines with time; hence a field that has 10 million barrels of oil reserves producing at 1 million barrels per year will not produce for 10 years and then just stop. Rather, the rate will decrease day by day and it may take 20 years or more before the last of the 10 million barrels is produced, sold, and the field declared uneconomic.

Just as Hubbert developed his theory during a time when oil was plentiful, so in the 1990s two geologists again warned that plentiful oil would run out sooner than many expected. Publishing in the popular science magazine *Scientific American*, Campbell and Laherrère (1998) pointed out that oil production would soon begin to decline. The popularity of the specific term "peak oil" is more recent, probably dating to press reports about the founding of an Association for the Study of Peak Oil in 2002[11] by, amongst others, Colin Campbell.

4.2 Campbell's Predictions of Oil Supply

While Hubbert's projections were founded in statistics, there are other ways to make a similar estimate. One of these is to use a data-intensive study to project oil well and oil field production declines. As discussed in Chap. 2, when an oil field is first discovered, the next step is to drill additional wells to find out its extent. As additional wells are drilled the production from the field will increase. But eventually the limits of the field are reached, the volume of the reservoir can be calculated with increasing accuracy, and oil is produced. As the oil is produced, the output from individual wells declines over time; the total output from the field, which is just the sum of all the wells, also declines. Every field is different, and such declines for a field may not start for a number of years as new wells are drilled to bring all areas of the field into production. Once declines start they may be slow and steady, or they may be precipitous. To some extent, the rate at which a well is allowed to flow determines the amount of oil that can eventually be produced from it, so production engineering decisions make a difference in the amount of oil produced. Periodically, it may be worthwhile to add new production technology, creating an upward step in the decline curve. In the 1990s, international exploration geologist Colin Campbell, working with a large, proprietary database,[12] summed up these patterns for all the world's oil fields. Campbell (1997) concluded that global

[11]See Aleklett (2012) p. 10 for a history of the term.

[12]Petroconsultants collected and compiled data from the exploration and production sector of the oil industry for much of the world, and then sold the compilations back to the oil companies. The company has since been acquired by IHS, which continues the activity.

oil production was about to become seriously constrained. This was the methodology which Campbell and Laherrère used in the *Scientific American* paper; their basic message was that as these declines set in on a global scale it would not be possible to reverse the trend by either new technology or new discoveries. That their total figures from a different methodology were similar to Hubbert's final predictions in 1982 lent credence to both approaches. Campbell (2013) has since published a detailed atlas of this work.

In the rest of this chapter the Hubbert approach, which uses global data to make predictions, will be referred to as a "top-down" approach; Campbell's summing up of thousands of individual fields methodology will be referred to as a "bottom-up" approach.

All theories have some underlying assumptions, and peak oil is no exception. Hubbert's use of a logistic curve to model oil production is one such assumption. Hubbert himself discussed the possibility of more than one peak, but Hubbert never clearly explained his decision to use a logistic curve. Deffeyes (2001), who knew Hubbert personally, opines that it was partly a lucky guess that seemed to work. Whatever the reason for choosing a logistic curve, the mathematical transformation to a linear plot allows calculation of the final amount that will be recovered when the last barrel is produced; Hubbert, and many others, concluded that this final number for oil is about 2 trillion barrels (2×10^{12} barrels). From that number, a knowledge of known production to date and symmetry the peak can be found. Deffeyes (2005) gives the date for this peak as being in November, 2005. The fact that production has increased since that date is clear evidence that the theory needs some modification, but why, and to what extent, remains hotly debated. In contrast, Hallock et al. (2014) show good agreement with a Hubbert analysis for many countries. Bentley (2002, 2009, 2016) gives a detailed analysis of the problems.

Despite some caveats in some of Hubbert's later papers, the symmetric form of Hubbert's curve has continued to influence the discussion. While there were and are many critics of the total resource figure, which is the area under the curve, and the date of the peak, the shape of the curve after the peak has received less attention. The logistic curve's symmetry suggests that just as the economy has phased into oil, it will phase itself out of oil without major economic upheavals. Thus the use of the logistic curve has, perhaps ironically, worked to lull concerns given that it shows a gradual replacement of oil energy rather than a crisis.

4.3 Supply Peaks Versus Demand Peaks

The gentle decline in production argues for a "demand peak" rather than a "supply peak." Hubbert, and most other geologists and oil industry analysts, start with a focus on the oil that is in the ground. Since this must be limited, their peak production is constrained by the supply of oil—it is a "supply peak." But the peak could be caused by some better, cheaper replacement for oil lowering demand. This is a "demand peak." One example of a demand peak is the production of anthracite coal in Pennsylvania (Carmalt and St. John 1986), as shown in Fig. 4.3.

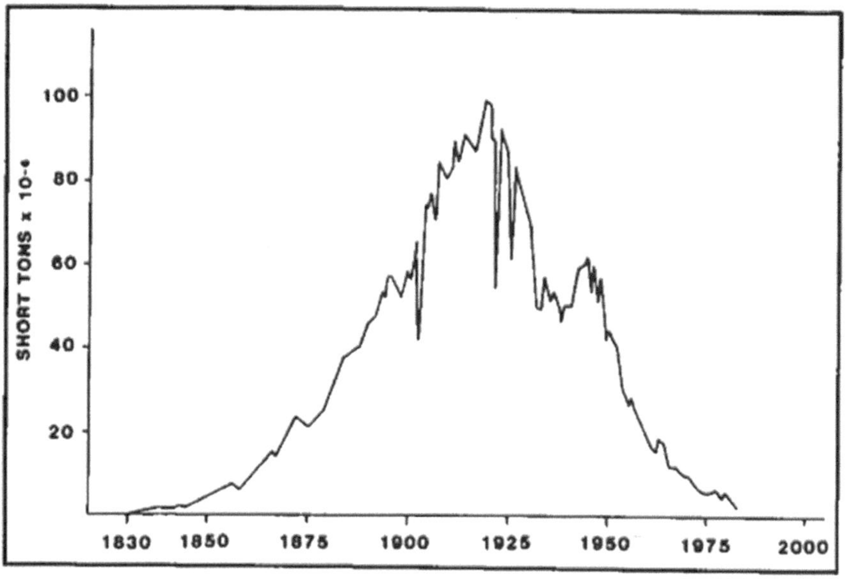

Fig. 4.3 Anthracite coal production in Pennsylvania (*Source* Carmalt and St. John (1986) © AAPG; reprinted with permission of the AAPG, whose permission is required for further use)

Northeastern Pennsylvania has not run out of anthracite coal; rather the cost of bituminous coal is so much lower that it is no longer profitable to mine anthracite. Economists generally like the demand peak approach, because it fits well with neoclassical economic theory. Those arguing for an oil demand peak say that oil will be replaced by some alternative energy when the alternative energy becomes less expensive; this will lead to the demand for oil dropping. The reason for some alternative energy becoming less expensive is presumed to be due to improvements in technology, but it could also be due to oil becoming more expensive due to scarcity. To the economist it is only the relative expense that matters, whether the alternative energy is another fossil fuel, such as natural gas, or some renewable energy technology. Given the ubiquitousness and importance of oil, and the magnitude of its use, it's a bit hard to conjure up a replacement.

4.4 Definitions in the Peak Oil Analysis

Both the top-down and the bottom-up approaches have had to contend with the fact that oil production has not clearly peaked.[13] Those who believe that an oil peak will be caused by supply limitations tend to argue that the reason is that

[13]At least it hadn't as of the end of 2015.

"unconventional" sources, such as tight oil and tar sand, and even natural gas liquids, have been added to the oil production statistics. By this standard, peak oil has been reached for the conventional oil fields that Hubbert analyzed, and the question becomes a narrower one of how long the rates of such unconventional production can compensate for the decline in conventional production. This has more impact on the bottom-up analysis than on the top-down analysis. In the bottom-up approach, one is always limited by the current state of the technology. By contrast, the top-down approach uses trends from historic data which have seen technologic improvement since the first wells were drilled in the nineteenth century; hence the projections into the future contain at least some embedded presumption of technological improvement. Foreseeing some of this, Hubbert mentioned the tar sands and oil shales in his papers, but there was no production from either of them when he was writing, so his statistics and resource estimates do not include them. Tight oil reservoirs were also known, but any oil they contained was not included in reserves, because until recently it was not thought that such low permeability reservoirs would ever be economic. Nevertheless, the result of adding tar sand production, which requires processing to become oil, to the oil production statistics does beg the question as to whether the data underlying the analyses are consistent. As mentioned earlier, oil shale is not presently being produced.

In addition to tight oil and tar sand, the unconventional label is frequently used to include offshore prospects in more than 500 m of water and prospects in the Arctic, particularly offshore in the Arctic Ocean. To date the Antarctic treaty, which prohibits commercial exploitation of that continent, has not been challenged. But when oil production peaks from non-Antarctic sources, do not bet that this treaty will not be challenged before much longer. Unlike tight oil and gas (especially shale oil and gas), tar sand, and oil shale, such prospects are unconventional only in the sense that they are pushing the limits of operations technologies, not in the sense that they are tapping a different sort of oil deposit. Exploration that occurs in these areas is traditional—the companies look for a good source rock which has reached "oil window" temperatures and from which the oil has then migrated into a porous reservoir where it is trapped.

Another source of additional oil from improving technology is enhancement of the percentage extracted from the reservoirs that are being produced today. Petroleum engineers know that the reserves for almost every field increase as time goes on. There are several reasons for this. First, the P1 reserves, which are the ones that can be listed as assets on a balance sheet, are necessarily conservatively calculated. Because they are assets they can be used as security for bank loans, and banks want to make sure of getting their money back. Frequently the bank will use lower price projections than the oil company to calculate the amount of security being proffered by the oil company. So the reserves shown are conservatively stated, and as the field produces, this figure can be revised upwards. Second, all the oil does not all come out of the rock. Depending on the rock's physical characteristics, only a fraction will flow into the well to be produced. As this natural flow declines, the company will use different techniques to coax more oil out from the rock. Such techniques are collectively known as Enhanced Oil Recovery (EOR) methods.

There are different methods, but many involve using some wells to pump something down into the reservoir. As this fluid enters the reservoir rock, it will push the oil in front of it into the producing well. Frequently this physical push will be combined with something that makes the oil flow more easily, for example steam, or a chemical that will mix with the oil and help it to flow to the producing well. These enhanced methods can follow one another during the course of producing a field and are thus called secondary recovery or tertiary recovery. But it is difficult to get 100% of the oil in a reservoir out; recovery can range from 20% of the oil originally in place in the reservoir to over 50%. The average is some 36%, and enhanced recovery adds only a few percentages at a rather high price.

Two comments about these enhancement technologies: first is that they cost money. Each such project will go through an approval process similar to the project financing example discussed in Chap. 2. The cost of drilling the injection wells, the costs of whatever is pumped into them, the possible cost of separating oil from this material when it shows up in the producing well, the time-series price of all of these materials and the price of oil over the project life all have to be worked out. This enhancement project will be another possible use of the company's capital, so it will be compared to other possible projects. The second is that it is unclear how much the additional production changes either Hubbert's top-down, or Campbell's bottom-up approaches used in forecasting peak oil. For the statistical approach, water was being used for secondary recovery in the Bradford giant oil field in Pennsylvania starting in the 1890s. For the bottom-up approach, secondary and tertiary recovery programs are the norm in the industry today, so the projections that Campbell and Laherrère use have included this recovery.

More fundamentally, the assumptions behind the peak oil projections are what an engineer would term the "boundary conditions" for the theory. All the conventional analyses presume the migration from a source rock into a porous reservoir rock from which the oil is produced by traditional means. Tar sands, which are mined (and which Hubbert mentions in his papers, but not in his analysis) do not fit these conditions. Shale oil and gas do not fit this condition; other tight plays are more similar to traditional plays but at the very lowest limits for permeability of traditional fields, and hence are at the limits of the boundary conditions. The result is that the projections based on any of the peak oil methodologies are likely to diverge from actual data. Indeed, this is the reason that production since about 2005 has not followed the peak oil projections.

As mentioned above in discussing demand-side peaks, economists have generally been scornful of peak oil theories. As Bardi (2013) has pointed out, the debate between those who see depletion as an urgent, near-term issue, and those who are sure that the combination of technology and human ingenuity make depletion a problem only in the distant future has been going on for well over a century. One of the first to consider the issue was the economist William Stanley Jevons in his classic work *The Coal Question* (1865), which considered what is essentially the same problem of resource availability for England's coal resources. Jevons was also an astute economist; he observed not only that the amount of coal had to be finite, but that the more that was mined, the more expensive the coal became. When he considered using coal more efficiently he noted "Jevon's paradox."[14] Basically the paradox is that using coal more efficiently does not reduce the amount of coal used; one might think that the more efficient use would result in less coal being needed, but in practice the use of coal increases because the more efficient use of coal effectively lowers its cost when measured per unit of energy. With the lower energy cost, demand increases, canceling out the efficiency savings. Jevons was able to demonstrate this with respect to coal in nineteenth century England. Modern environmentalists know this as the "rebound effect" and have observed it in modern situations (Knittel 2011).

The reason that economists like the demand-side explanation for peak oil is that it better fits their economic framework. If there is a shortage, the price will rise and more of the material will be produced. Skinner (1986) provides a good review of how this works in the mining industry. The late Morris Adelman, an MIT economics professor, is as strongly identified with the position that increasing demand will bring forth increasing supply (Adelman 1993) as Hubbert is with the position that there are geologic limits to the amount of oil. This approach can be seen in the work of the US Geological Survey. The USGS is a respected scientific organization which is apolitical and was established "... to examine the geological structure, mineral resources, and products of the national domain" (Rabbit 2000). In 2000 the survey published an extensive assessment of world petroleum resources, arriving at a final estimate of 3×10^{12} barrels of likely recoverable oil, or about half again as much as most of the peak oil estimates (USGS 2000). One of the key concepts in the USGS methodology is that of the "resource pyramid," as described by McCabe (1998) and shown in Fig. 4.4.

The concept behind the pyramid is that at any given time, the combination of price and technology can be represented by a horizontal plane through the pyramid. Above the plane lie the current economic resources, and below the plane lie those that are uneconomic. Hence as oil becomes more scarce, the price rises and the plane moves to a lower level. This will make a much greater part of the resource available as Adelman predicts.

The pyramid approach to a finite resource is similar to that of decreasing ore grades known from mining. Skinner (1976) has argued that for some metals, the distribution of the metal in the earth's crust is bimodal. For example, lead can be found dispersed in granite, but the amount of lead is so small, just 0.0039% in an average granite, that it would not be economic to mine granite in order to obtain lead. Rather, one looks for special geological situations in which lead forms the mineral galena (PbS), which may comprise about 10% of the rock in a lead mine. It is this concentration of lead molecules into a mineral that makes an ore, and hence a lead mine.

[14]See Owen (2010) for a precis with commentary of Jevon's paradox.

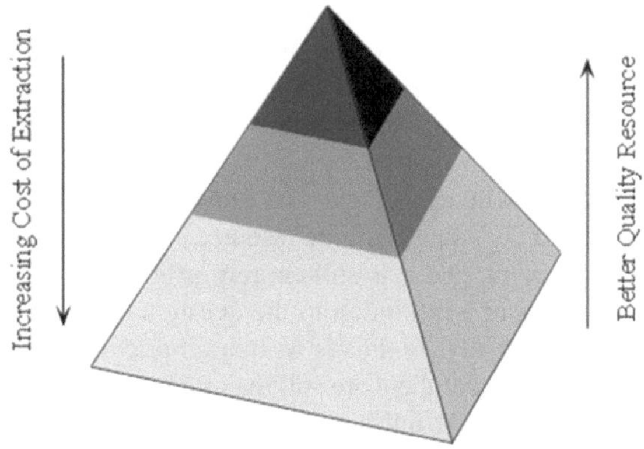

Fig. 4.4 Resource pyramid (McCabe 1998 © AAPG; reprinted with permission of the AAPG, whose permission is required for further use)

Both the bottom-up and the top-down oil methodologies agree that most of the oil reserves, at least to date, are found in giant oil fields. Giants are fields that have more than 500 million barrels of recoverable oil. The most recent compilation (Horn 2011) indicates that approximately three-quarters of the projected global resource is contained in the giant oil fields. Perhaps the giant fields should be viewed as similar to ore minerals, and hydrocarbons further down the pyramid should be considered more like the lead distributed in granite. But unlike the dispersed lead in granite, some of the dispersed hydrocarbons—notably the tight oil deposits like the Bakken and the tar sands—are marginally economic at present.

Thus many in the oil industry do not see any immediate constraint on oil supplies, even if the conventional peak oil is behind us. For example, the IEA (Tanaka 2009) suggested that the total amount of oil available to the world economy at prices which are within reach of affordability is vast.

Two things are clear. First is that definitional problems abound. While the BP series showing reserves shows an increase from year to year, the details in the data show that Canada increased its oil reserves almost fourfold from 1998 to 1999 as the result of a decision to include the tar sands in the numbers; similarly, inclusion of the Orinoco tar sands in Venezuela's reserves resulted in that country showing an almost twofold increase from 2007 to 2008. In both cases the resources were well-known for many years before they were considered in the reserve numbers and the inclusion was due simply to resources moving from the resource to the reserve category. Second is that across all countries, the higher oil prices of the twenty first century (until 2014) led to a transformation of resources into reserves.

Since Hubbert's original 1956 paper the two things that everyone wants to know are: how much oil is there? and when will it be impossible for oil production to increase? The debate that Hubbert joined in 1956 continues today. In BP's annual statistical tables, 2015 is the first year that shows a slight decrease in oil reserves. But until there are several more years of data it would not be clear whether this decrease is the result of the low oil prices of 2015 requiring reclassification of some reserves into just a resource, or whether for the first time discoveries were insufficient to replace production.

There are a number of important points which are generally overlooked. The first is that, even if current oil production is at about its eventual peak there are sufficient unconventional oil sources to meet immediate needs, albeit at higher prices. What is not so clear is whether the needed investment is being made to use such resources over the intermediate term, and how to address the long-term problem that exponential growth cannot ever be supplied with a resource which does not grow exponentially—and no finite resource can grow exponentially. That was Hubbert's original point, and it is still entirely valid.

So is there a resolution to the question of whether we have reached peak oil in 2016? Not really. As this is written oil prices have increased somewhat since their lows in 2015, but there are still major concerns about whether most new oil projects will be economic unless oil prices continue to increase. The activity and developments that these higher prices encourage result in very expensive oil (low EROI values), which tends to indicate that the peak of traditional production has been reached. But the price fluctuations of the past two years also indicate that more than just simple geology must be included in order to answer this question, which is what the next chapters will explore.

Chapter 5
Energy in the Economy

Energy is not like other resources. Whether mined as coal, produced from wells as oil or natural gas, released in a nuclear reactor, or captured by panels or windmills, energy is consumed. More precisely the energy—which we all learned is neither created nor destroyed—is changed from a more concentrated form to a more diffuse form. The fossil fuels contain energy that has been chemically stored; when fossil fuels are burned this chemical energy is converted into heat energy.

5.1 Some Basic Thermodynamics

Studying the flow of energy is the science of thermodynamics, literally "heat flow," which lies at the heart of all dynamic systems.[1] The earth, its oceans, ice caps, rivers, ecosystems, plants, and animals are all dependent on flows of energy. So, too, is "the economy," which is what allows us to have a style of living so very different from our hunter-gatherer ancestors. Particularly important to the way energy, and thus oil, moves through the economy are the First and Second Laws of Thermodynamics. The First Law, also called the Law of Conservation of Energy, states that energy is neither created nor destroyed, but only is transformed from one form into another. As modern science developed in the seventeenth and eighteenth centuries, the conservation of various forms of energy became clearer. But it was not immediately obvious that the different forms of energy were all, in fact, the same physical entity. In particular, the equivalence of heat and mechanical energy was only established in the late eighteenth century, in part by being able to observe the conversion from one to another during the manufacture of canons.

[1]The subject of thermodynamics is covered in many textbooks and online resources. Principle references consulted for this discussion are: Gallucci (1973) and Larsen (2014).

S.W. Carmalt, *The Economics of Oil*, SpringerBriefs in Energy,
DOI 10.1007/978-3-319-47819-7_5

The Second Law of thermodynamics involves entropy and states that entropy cannot decrease.[2] What is entropy? It is the amount of **unuseable** energy contained in a substance at a particular temperature. More simply, the Second Law says that the amount of useable energy can only decrease.

This is why we think of energy as being consumed, despite the First Law stating that it is not. When we "use" energy, we are converting it from a usable form to an unusable form. The form which cannot be used is generally heat, which we release into the environment. If I am in a large room in which there is a cup of hot water, the hot water will cool as heat flows from the cup into the rest of the room. At the start, the heat of the water could poach an egg; after the cup of water cools it cannot. The total heat in the room has not changed, but it has dissipated into a much larger space, which makes it unuseable. The energy before and after is the same (First Law), but while before there was some useable energy, after there is only unuseable energy (Second Law). Because there is more unuseable energy afterwards, the entropy has increased. Within the room we cannot get that energy back to a useable form again. Because entropy can never decrease overall, it is sometimes called "time's arrow."

The example of the cup of water in a room also illustrates two additional points about thermodynamics. The first of these is that our thought experiment of the hot cup of water in a room involves a flow of heat until the heat is evenly distributed in the room and no more heat flows. This is an equilibrium condition. The second point is that a system can only reach equilibrium if we have boundaries that prevent anything from entering or leaving. In the almost two centuries since the basic principles of thermodynamics were first worked out, thermodynamic analysis has proved to be a robust explanation of physical and chemical processes.

When thermodynamics is considered at a molecular level, entropy is found to be associated with the amount of disorder amongst the atoms and molecules of the substance or system being studied. Saying that entropy is a measure of disorder is thus a useful, although perhaps oversimplified, way of understanding the concept of entropy. This is illustrated in Fig. 5.1.

Is there any way to make entropy decrease? There is, and it happens all the time. The hot cup of water is an example. How did it get hot? The short answer is that we added heat, i.e., energy, to it. On analysis, everything that we do which requires energy is creating a local area of lower entropy at the expense of increasing the entropy of another system; this local low-entropy area will dissipate over time, increasing in entropy as its energy diffuses into the surroundings and becomes unuseable.

[2]We generally learn the second law very early in life; the nursery rhyme Humpty Dumpty is one articulation of the Second Law.

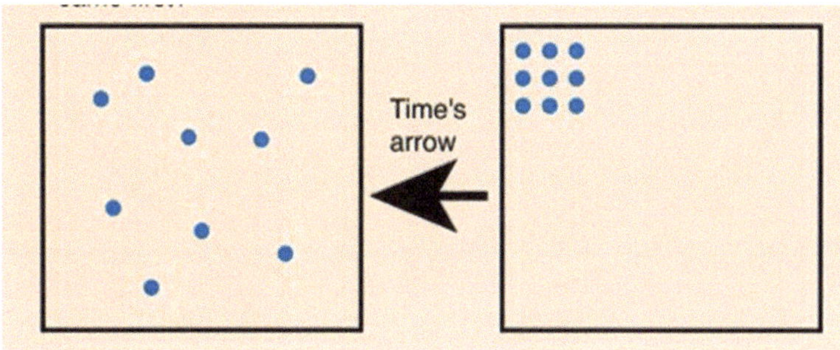

Fig. 5.1 Ordered and disordered molecules in box (Nave 2000)

A short digression: Terms used for thermodynamic systems

Thermodynamics is a **macroscopic** description of changes. The part of the universe that is being examined is the **system**, everything else is the **surroundings**. A system is separated from its surroundings by boundaries.

An **isolated system** has neither energy nor matter crossing its boundaries.

A **closed system** allows energy across its boundaries, but not matter.

An **open system** allows both matter and energy to flow in or out.

A system is at **equilibrium** when there are no more changes occurring. This is easy to understand for an isolated system.

Closed and open systems can also be in a condition of **steady state**, which is when a constant flow of energy is required to maintain other properties in an unchanging state. For example, the steady state of a refrigerator at 5 °C is possible only by the energy used to keep it cool. To avoid confusing a steady-state condition with an equilibrium condition, examine whether a flow of energy is needed to prevent any change.

A **reversible** reaction is one which, as the word implies, can be reversed. For example, in a closed system we observe that ice melts when the temperature rises, and then re-freezes when the temperature falls.

An **irreversible** reaction is one which will not (at least simply) go in the opposite direction. For example, if one puts a teaspoon of solid salt crystals into a swimming pool they will dissolve; no simple change to the water containing the dissolved salt will result in a teaspoon of solid crystals.

Note that our thermodynamic example of the cup of hot water in a room has a time dimension only until the water and the room are at the same temperature, that is in equilibrium. Once at equilibrium the system is time-independent.[3]

[3]That the equilibrium will never change in our room is not quite true—over billions of years the sun's nuclear reactor will use up all its fuel and the solar system will cool down to a temperature very close to absolute zero—increasing the universe's entropy as it does so.

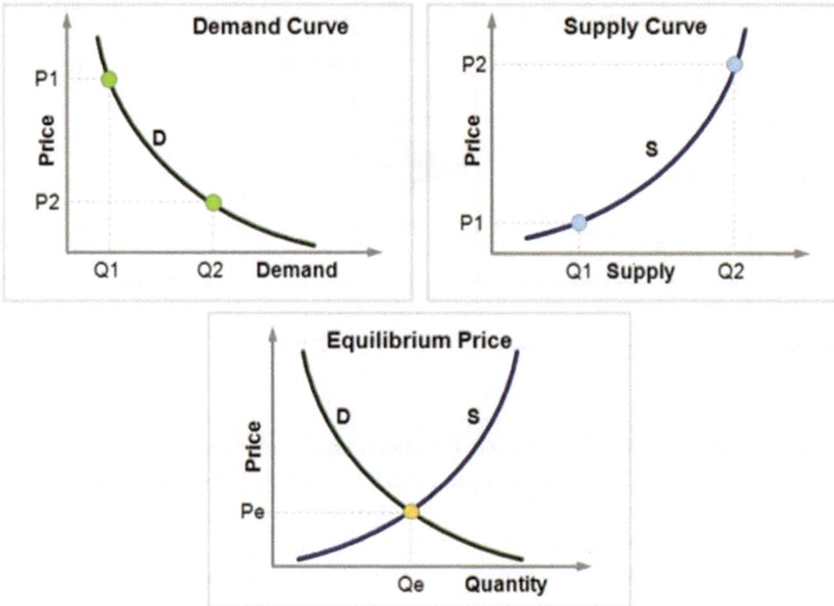

Fig. 5.2 Demand, supply, and the determination of price as the steady state between the two (Rodrigue 2013)

Parallel to the development of the laws of thermodynamics in the nineteenth century, economists were also looking at equilibrium. A foundation concept, taught in elementary economics, is that both demand and supply vary continuously with respect to both price and quantity, and these two find their equilibrium where they meet, as shown in Fig. 5.2. While economists consider this an equilibrium, it is actually a steady-state condition.

In the later half of the nineteenth century, economists developed such equilibrium models using the same mathematical tools that the physicists were using for energy equilibrium studies. Generally called "neoclassical economics" this equilibrium approach to understanding the economy has remained central to much of economic theory to the present day. Quite self-consciously the three economists[4] who, essentially independently, developed the neoclassical approach were attempting to make economics a science that was just as respected as physics. Walras (1874) was particularly keen to be able to bring the mathematical rigor of physics to economic problems (Beinhocker 2006), although Jevons had already published his book, *A General Mathematical Theory of Political Economy*, in 1862. Whether in physics, chemistry or economics the equilibrium state is the simplest, and hence the mathematics of equilibrium is the most tractable.

[4]The three were Jevons in England, Walras in France and Menger in Austria.

But the natural world is not in equilibrium. What may seem at first glance to be at equilibrium is generally a steady state. An analogy is the human body. An adult human may look the same day after day or even year after year, but he or she is not at equilibrium with the environment. Rather, he or she is continually processing flows of air, food (chemically stored energy), and water. Similarly, an entire ecosystem is supplied with energy (primarily sunlight), nutrients, and water, all of which flow through the system.[5] Descriptions of equilibrium are useful because they indicate the direction in which reactions will occur. But it is only a direction; in a steady state the equilibrium condition is never reached.

5.2 Energy Systems

The economy is similar. It is a dynamic system which is continually processing various inputs to provide the outputs which we value and measure as GDP. When one thinks about any economic activity—mining, shipping, cooking, working, entertainment, think of anything—a short reflection will confirm that the activity uses energy. Even the financial markets, which at first glance might appear to be devoid of an energy input, require energy. Financial markets require data, which is then processed. Collecting the data takes energy, even if only small amounts of electricity to directly store a data point in a computer. In addition to the energy that is used directly, there is the energy that various humans use in collecting and processing such data, and in interpretation and decision-making. In short, every economic activity requires energy.

The flow of energy through a system results in an increase in entropy. That is why we get hot and sweat when we exercise. Our bodies maintain a balance between the energy we take in and the less useable energy we return to the environment. Similarly, the earth maintains a balance between the high-grade energy that it receives from the sun and the low-grade heat energy that it re-emits into outer space. Unlike the closed equilibrium system, the output of an open steady-state system will need just as much energy today as it did yesterday, and it will need the same amount again tomorrow. This applies to the global economy just as surely as it does to the human body or to the entire earth. If the economy is growing, it will need energy to maintaining its steady-state condition as well as additional amounts for the growth. Because of the Second Law, all of this energy use will become unavailable and there will be an increase in entropy.

The economist Nicholas Georgescu-Roegen understood that the economy was a steady state rather than an equilibrium system. His 1971 book *The Entropy Law and the Economic Process* represents a completely different view of economics from that of the neoclassical or Keynesian economic models which focus on equilibrium

[5]See Hall and Klitgaard (2012) for extensive discussion of energy flows through ecosystems.

economic states.[6] Georgescu-Roegen's view of the economic process, while still considered iconoclastic by most economists, held that value is created by irreversible thermodynamic processes, i.e., by processes that increase entropy. In examining the economic process from a thermodynamic perspective, Georgescu-Roegen argues that not every process that creates entropy is an economic one; rather, to create economic value, the process must be irreversible, create a local area of low entropy, and that the process must create something that human beings value.[7]

5.3 Ecologic Systems

Seeing the economy as a steady-state system rather than as an equilibrium system is a major leap in outlook for many economists. The viewpoint may be easier to grasp by analogy by looking at some of the concepts in ecological systems that have been developed over the past half-century.[8] Photosynthesis is used by plants to live; plants process substances in their environment, generally minerals from the soil and carbon in the form of carbon dioxide from the atmosphere, with energy from the sun to create a low-entropy organism. We can see that the plant is a low-entropy island when it dies, because it decomposes. Herbivores eat plants, and carnivores eat herbivores: in each case the animal uses its food as a source of energy and converts this to provide maintenance of itself as a living organism. At each point in the food chain, the consumer is a low-entropy island, and maintains this low-entropy state by the energy embedded in the food eaten. Furthermore, each step of the process creates some entropy, so the total amount of energy needed increases the further up the food chain one goes. When its energy supply vanishes, the organism can no longer survive.

It is fairly easy to show that pollution in the environment and entropy are linked. Food is a source of energy, but, uneaten, it becomes garbage and begins to decompose. Like salt dissolving in water or smoke from a chimney, such decomposition disperses energy into the atmosphere. Georgescu-Roegen's ideas are thus applauded by most environmentalists and ecologists. His reception amongst his economic peers has been more uneven. Most of the economic arguments made against Georgescu-Roegen's approach are based in the neoclassical world view that, as we have noted, views the economy as an equilibrium system rather than as a steady-state system.

[6]Georgescu-Roegen (1971).

[7]Beinhocker (2006) gives a synopsis of Georgescu-Roegen's ideas.

[8]There are many texts in ecology. For greater detail see *Fundamentals of Ecology* (5th edition; E. Odum and Barrett 2005) and *The Systems View of Life* (Capra and Luisi 2014) which give much more complete treatments of the subject.

5.4 Measuring the Economy

Whatever the outlook, we need ways to measure the economy. The usual bulk measure is its Gross Domestic Product (GDP), for which a succinct working definition is provided by the government agency that measures it for the USA. According to the BEA,[9] the GDP is "the output of goods and services produced by labor and property located in the United States."[10] There are a number of problems with this definition. For example, should we include the labor involved in services such as child care or care for elderly parents as opposed to only counting labor when we pay for such services? Another problematic value, large in the case of the USA, is the "imputed rent" that the owner of a home does not have to pay in cash in order to live in a home that he or she owns[11] Note that both armament manufacture and health care for wounded soldiers are positive additions to GDP, despite a nagging suspicion that they should somehow have opposite signs; the same problem exists for the costs of environmental cleanups and the industrial activity that resulted in the environmental problem in the first place.

The fact that the cost of an environmental cleanup and the activity which caused the need for the cleanup both contribute to GDP is because both use energy. In both cases, the use of energy results in an entropy increase. This example simply confirms that the economy can be viewed as a flow of energy.

Combining GDP from different countries leads to questions about how to include trade and what exchange rates to use (e.g., nominal, purchasing power parity), particularly in countries which have fixed exchange rates. Finally, comparisons of GDP from year to year raise additional questions of how the underlying money may have changed from year to year. Adjusting the values by some inflation index is what is usually done, but such indexes and adjustments are just as problematic as is the single GDP calculation. Furthermore, any problem with an inflation index tends to compound, similar to compound interest. So the GDP numbers that are widely published in the financial news give a sense of what is happening. But they are not precise measurements and should be taken as but a rough indication of economic activity, especially when viewed over a length of time. Still, for all its faults, GDP is the best number available, so it is what gets used.

[9]Bureau of Economic Analysis of the US Department of Commerce.

[10]The definition is as given in the BEA news release for the 2014 2nd quarter GDP figure at http://bea.gov/newsreleases/national/GDP/GDPnewsrelease.htm accessed 2014-09-10.

[11]In the USA imputed rent is included in the GDP calculation.

5.5 Energy and the Economy

We are now in a position to examine more readily the relationship between energy use and GDP. As we do this, we need to keep in mind that many economic principles will only indicate the direction in which changes will occur, given that an equilibrium state will not be reached.

The first relationship between energy and GDP is that they are correlated; higher global energy production is associated with higher global GDP. This is to be expected, given that the economy (GDP) is a steady state at any given time. A small economy at a small steady state will use less energy than a large economy at a large steady state, all other things being equal. So as the economy grows from one steady state to the next, so, too, will the amount of energy used grow. This relationship is shown in Fig. 5.3.

Where the lines are not exactly parallel it means that the energy is being used more or less efficiently in the maintenance of the steady state. So Fig. 5.3 shows that, since 2000, there has been some improvement in overall energy efficiency, i.e., more GDP per unit of energy. The ratio of energy to GDP is more properly referred to as the "energy intensity."

Figure 5.4 shows these same relationships separately for OECD and non-OECD countries. Here we see that the intensity gain has been entirely in the OECD (i.e., developed) economies (solid lines). In the developed economies, the GDP has continued to increase since 2000 whereas Primary Energy in these economies has been almost flat over this period. But this improved Energy Intensity in the OECD countries is offset by a worsening trend in the non-OECD countries (dashed lines) where the Primary Energy use is increasing at least as fast as the GDP.

While the OECD countries are prone to boast about their recent increased energy intensities, the above trends can also be used to indicate that at least some of the OECD improvement has been due to selective transfer of energy-intensive activities to non-OECD countries.

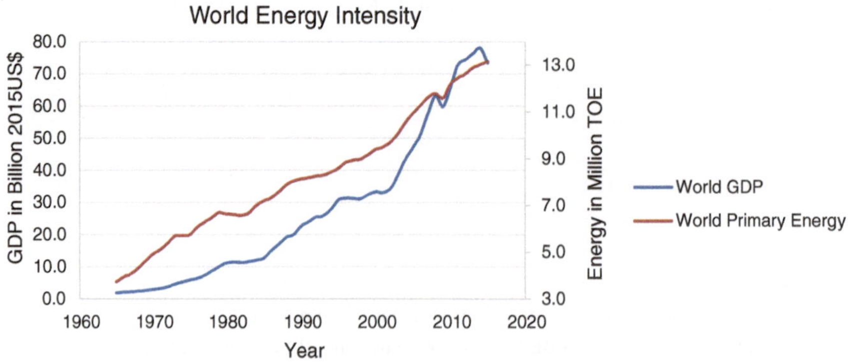

Fig. 5.3 World energy intensity (Data sources: World Bank 2016, and BP Statistics 2016)

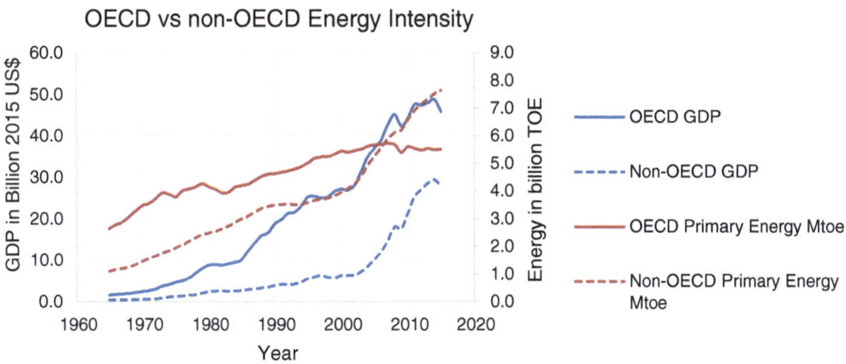

Fig. 5.4 Energy intensity of OECD (*solid lines*) and non-OECD (*dashed lines*) (Data sources: World Bank 2016, and BP Statistics 2016)

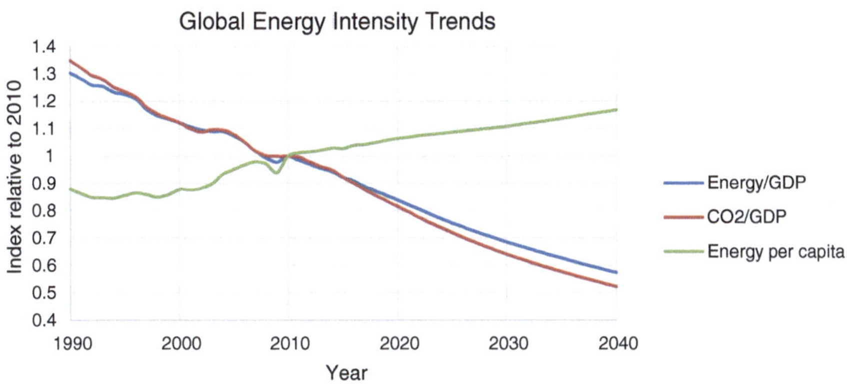

Fig. 5.5 Global energy intensity trends (Data sources: EIA 2016; United Nations 2015)

Globalization, which is a combination of trade rules, good communications, and improved transportation, defines an open state system. This requires that measures of energy intensity be viewed from a global perspective; hence pollution and other entropy increases must also be viewed from a global perspective.

Figure 5.5 shows three different calculations of energy intensity for the world: the energy used per unit of GDP; the CO_2 emitted per unit of GDP; and the energy used per capita. Because most energy in the economy is obtained by burning fossil fuels, comparing the energy and CO_2 with respect to GDP are very similar. The increase in energy per capita reflects improved living standards. But ratios hide absolute values. Decreasing energy or CO_2 intensity can be due to increasing GDP rather than decreasing energy use or emissions. Hence Fig. 5.5 is not saying that the absolute amounts of energy used and CO_2 emissions in 2040 will be less than at present.

Another relationship between energy and GDP can be seen in the relationship between the energy price and the GDP. From the classical economic supply and

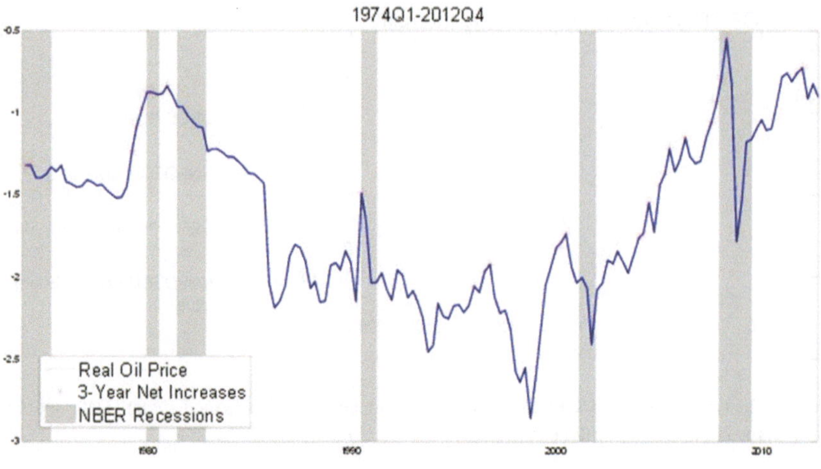

Fig. 5.6 Oil price increases and recessions (Kilian and Vigfusson 2014)

demand model, one expects that when the price goes up the demand will go down and the supply will go up until a new equilibrium is reached. But thinking back to our economic models, there is likely to be a significant time delay until the new supplies become available. The immediate effect will be that there will be less energy sold at the higher price. And because energy and GDP are correlated, the price increase will result in a lower GDP—a recession.

The International Monetary Fund (IMF) has plotted the oil price and economic recessions, which shows that recessions are frequently associated with high oil prices either just before or at the start of a recession. In particular, the price increases associated with the fall of the Shah of Iran (1979) and the Gulf War (1990–1991) can be seen in Fig. 5.6, which plots data from 1974 to 2012.

From 1971 to 1980 oil prices rose from $3.60 to $37.40 per barrel, an increase of over 10 times. The economy did adjust to this increase, but only slowly. Cause and effect can be difficult to sort out; monetary and credit policy (Kilian and Vigfusson 2014) also are important. Whatever the cause(s) of a recession, there are additional complications with respect to the oil markets. For example, the surge in exploration that resulted from high prices during the 1970s led to a glut of production in the 1980s. The high oil prices of the 1970s also appear to have permanently changed the trajectory of peak oil, as shown in Fig. 5.7.

The problem for economic planners, and therefore for politicians, is twofold: first is the fact that as the finite resource of oil gets consumed, the marginal cost of new supplies goes up, and second is the fact that the project time-frames for energy projects make any rapid adjustment to a new situation difficult at best.

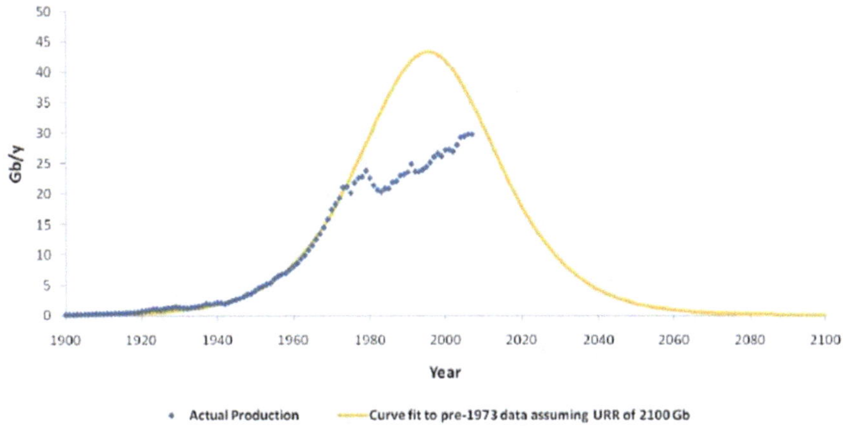

Fig. 5.7 Oil production curve (Bentley et al. 2009)

5.6 Net Energy and EROI

Hubbert, of the peak oil theory, pointed out that no company will produce a barrel of oil if more than a barrel is consumed to produce it (Hubbert 1982). This is, in essence, the definition of "net energy" (Peet and Baines 1986)

$$\text{Net Energy} = \text{Energy Produced} - \text{Energy cost}.$$

To make net energy numbers comparable, the cost portion needs to include three components: the energy used in developing the energy source, the energy used in ongoing operations, and the energy needed for eventual decommissioning. The calculation is done in energy units, for oil generally in barrels. But net energy is not much used because there are many other measurements available. For example, rather than considering the net energy of oil exploration, it is far easier to use figures such as reserves discovered or average size of new discoveries.

But a ratio using the same data as is required for net energy allows comparisons that are not available by other means. The most common is the Energy Return on Investment (EROI),[12] developed by Charles Hall from the 1970s. The original concept was developed in ecological studies which describe the efficiency of trout in finding food to eat and being able to spawn the next generation.

The definition of EROI, analogous to the financial ROI discussed earlier, is

$$\text{EROI} = (\text{Energy Produced} - \text{Energy costs})/\text{Energy costs}$$

[12]Sometimes also termed EROEI (Energy return on energy invested) and EROIE (Energy return on invested energy).

which is to say

$$\text{EROI} = \text{Net Energy}/\text{Energy costs}$$

with energy costs comprising capital investment, operating costs and decommissioning costs, all in energy terms.[13] Using EROI we can compare projects across many different types of energy systems. Furthermore, because the EROI calculation is independent of time, calculation of EROIs for projects completed at different times can be compared directly, without having to adjust for economic changes.

What matters is the Net Energy, that is to say the energy available to the overall economy. A comparison showing the percentage of Net Energy in relationship to the EROI is shown in Fig. 5.8.

As the graph shows, the difference in Net Energy between an EROI of 50 and an EROI of 20 is not huge. A project with an EROI of 50 means that of 100 energy units produced, almost 2 will be used in the production of the energy, leaving 98 for the wider economy. A project with an EROI of 20 means that the project will provide a net contribution of only about 95. An EROI of 10 indicates the project will contribute only about 90 units. Lower than 10, the drop-off in Net Energy becomes very rapid, hence this graph is frequently called the "energy cliff" by analysts; a project with an EROI of only 1 will be consuming as much energy as it produces, hence will not be contributing any energy to the wider economy. But note

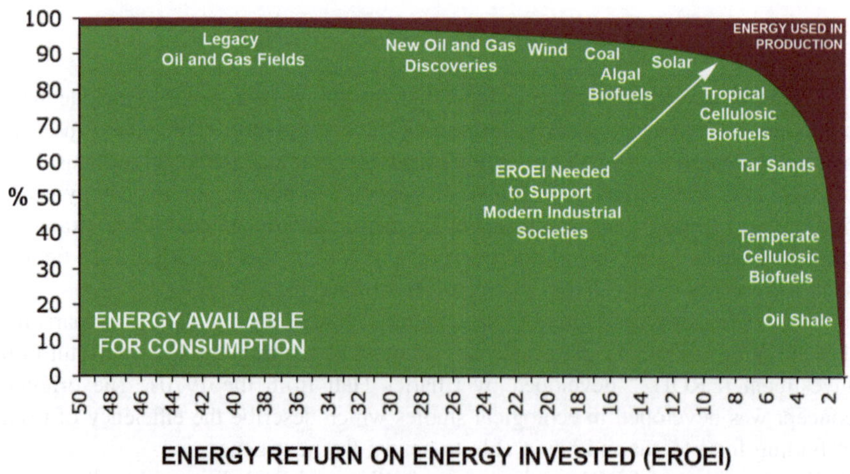

Fig. 5.8 The Net Energy "cliff" (after Mearns 2006; Murphy 2011 and Mearns 2016)

[13]Somewhat similar to EROI is the Energy Return Ratio (ERR). Defined as simply

$$\text{ERR} = \text{Energy produced}/\text{Energy costs},$$

the relationship between EROI and ERR is EROI = ERR − 1.

that the financial investments in an energy project with an EROI of 1 would still be counted as part of that country's GDP, which thus might be considered the "event horizon" of energy economics by analogy to the event horizon around black holes found in astronomy (see Blundell 2015).

The EROI, ERR and Net Energy numbers are always in energy units; the calculation is never done in monetary units. This means that financial ROIs and EROIs of a group of projects may rank in different orders. If an inexpensive, presumably historic, source of energy is used for the energy inputs the financial ROI may be significantly higher than the EROI. Some analysts argue that this is the current situation with respect to most of the "tight" oil and gas projects being undertaken today (e.g., Smith 2012). There is considerable discussion between economic analysts and energy analysts as to whether such differences between ROI and EROI can long be supported within an economy (e.g., Nelder 2012; King and Hall 2011).

5.7 Economics of Future Energy

In the last chapter, we discussed how any natural resource declines in quality the more it is used. This is another way in which oil and natural gas are different from, for example, copper. One might construct a CuROI (Cu being the symbol for copper), that is, the amount of copper needed to produced new copper. Or an AuROI, the amount of gold required in the production of gold. Just as for energy, these figures will become lower over time. But because of recycling, they do so relatively slowly (Bardi 2014). Because of the Second Law, we cannot recycle energy, and its EROI decreases faster. Over the course of the twentieth century it is estimated that the overall EROI of oil fell from over 100:1 to somewhere around 20:1.

The global economy is growing for three reasons: population growth, economic development in less-developed economies, and growth within most economies; growth in energy supplies is needed to support each of these. How the needed energy will continue to be supplied is one of the critical issues of the twenty-first century; more specifically, the problems are how to continue to provide energy for the economy with a decreasing EROI that already seems to be dangerously close to the "energy cliff." Remembering the Red Queen in Lewis Carroll's *Through The Looking Glass*, this problem is often referred to as the "Red Queen" effect—one must run faster and faster just to stay in the same place. As Garcia (2009) shows, it is possible to change the details of such projections in ways that appear to solve the problem, but the problem then simply reappears differently.

Projections[14] of energy supply and demand almost always examine the three issues mentioned: population growth, per capita energy usage growth and societal

[14]At least every projection of which I am aware that looks at total, global energy supply and demand. See IEA (2013), EIA (2014), the Russian Academy of Sciences (Mitrova 2013), ExxonMobile (2012), Royal Dutch Shell (2009).

Table 5.1 World sources of energy 2013

Energy source	Million tons oil equivalent[a]	Percent of total (%)	Average percentage growth last 5 years (%)
Oil	4185.1	31.9	0.9
Natural Gas	3020.4	23.1	2.0
Coal	3826.7	29.2	3.2
Total Fossil Fuels	11,032.2	84.2	2.0
Nuclear	593.2	4.5	−0.9
Hydropower	855.8	6.5	3.3
Wind	142.2	1.1	23.4
Biofuels	65.3	0.5	7.0
Solar	28.2	0.2	63.6
Geothermal, biomass, other	108.9	0.8	8.8
Other	279.3	2.2	17.7
Total renewables	623.9	4.8	16.2
Grand total	13,1058.1	100.0	2.4

Source BP (2014)

[a]Converting everything to "tons of oil equivalent" allows analysis, but has many methodological problems. For oil, average values of weight per barrel and energy per barrel are generally used to calculate the energy in a "ton of oil equivalent" (toe). But different sources use different average numbers. Similar to oil, average values are used to convert natural gas and coal to their energy content and then convert that energy using the average oil figures back to toe. It gets more complicated when there is no fuel being burned. The heat generated by a nuclear reactor is generally known; because this heat is used to make electricity in the same way as is heat from burning coal, the calculation of toe from heat value is still relatively straightforward. Renewables, including hydropower, where the applications and technologies are different pose additional problems. Typical thermal electricity generation has an efficiency of about 33%, so often the electricity output from such sources is divided by a figure of between 2.5 and 3.5 to arrive at the toe equivalent figure. But this presumption and division is only sometimes done, making comparisons tricky. See American Physical Society (2014) and Martinot et al. (2007) for a more complete discussion.

energy growth. The resulting GDP projection is correlated with energy consumption to give the total forecast of energy demand. Some studies create several projections according to different scenarios. For example, the IEA's 2013 *World Outlook* looks at three scenarios: a "current policy" scenario, sometimes also called "business as usual," which makes projections based on legislation, practices and policy currently in effect; a "new policies" scenario which make projections based on legislation, practices, and policies that have been announced by governments but are not yet implemented; and finally a "450 scenario" which presumes that governments will implement policies which will restrict atmospheric CO_2 to 450 ppm.

All of the various projections point to global GDP growth of between 3 and 4% over the coming decades. Oil is the critical component of the energy mix due to its role in transportation. As of today, there are no major alternative energy sources available for large-scale transportation. This poses three quite different problems:

the first is that it is not clear that such growth will be able to be maintained, given the decreasing EROI values for oil; the second is that if the declining EROI values do not result in energy shortages, it means that there will be far too much CO_2 released into the atmosphere to allow meeting the 450 ppm CO_2 limit that is the maximum that can be tolerated without run-away global warming, and finally, given the long capital investment lead times for energy projects, there are concerns that there simply is not sufficient time to make changes. Table 5.1 shows the recent (2013) distribution of energy sources.

If the GDP growth of between 3 and 4% is to be achieved, and the 450 ppm CO_2 limit also honored, then the nonfossil fuel components of total energy will need to go from being less than 20% of total energy at present to virtually all of it within two decades. The nuclear contribution will be limited in OECD countries due to political constraints, and sites for major new hydroelectric projects are rapidly declining. This leaves major contributions from other renewables as the way forward. While some of these have grown rapidly over the past 5 years, there is some concern as to whether these growth rates can be sustained. For comparison, mobile phone accounts, which over the past two decades have virtually exploded, have been growing at a rate of about 10% per year (Kearney 2013).

To sum up, unlike other resources that society uses, energy cannot be recycled. Furthermore, as with all resources, the energy sources that provide the best return are the ones that get used first. Population growth coupled with a rising standard of living points to global GDP growth over coming decades. The growing GDP requires more energy, while at the same time declining EROIs mean that the need for energy grows even faster than the GDP growth.

Chapter 6
Oil's Future Role in the Economy

The nineteenth century might be considered the century of coal. The harnessing of the coal's stored energy by development of the steam engine led to the industrial revolution and a concurrent revolution in transportation. Most of the fundamentals were in place by the middle of the century, and the continued rapid growth in the economy and in technology during the second half of the nineteenth century occurred within this context. In the second half of the century the world was familiar and people thought they understood it.[1]

In many ways, the twentieth century was a parallel to the nineteenth, except this time oil was the new energy source. During the first half of the twentieth century the new fuel replaced the old as the critical energy source. Although coal continued to play a major role in providing energy, it was oil that was at the cutting edge. Oil was particularly important in transportation where it revolutionized the way both people and goods move. And, just as in the nineteenth century, the second half of the twentieth century has seen continued technologic and economic development within the context set in the first half of the century.

This is not to say that other energy sources have not contributed throughout. As described in Chap. 5, oil represents less than 1/3 of the total energy supply. But it is a critical third of the energy; many would say ***the*** critical third. One of the significant developments in the global economy in the second half of the twentieth century has been the integration of national and regional economies into a single global economy, the foundation of which is the ability to move raw commodities, finished goods, and people efficiently and inexpensively from any one place to any other place. This transportation capability, from ocean ships to trains to cars and trucks to airplanes is fueled almost entirely by oil (Friedemann 2016). Referring to

[1]The great British physicist, Lord Kelvin (1824–1907), is reported to have said near the end of his life "There is nothing new to be discovered in physics now, all that remains is more and more precise measurement"; this accurately conveys the sentiment of the time despite the fact that the attribution to Lord Kelvin is probably incorrect.

S.W. Carmalt, *The Economics of Oil*, SpringerBriefs in Energy,
DOI 10.1007/978-3-319-47819-7_6

concerns about peak oil, Hirsch (2012) has observed "there is not an energy crisis; there is a liquid fuel crisis".

It is already obvious that the first half of the twenty-first century is again going to see a major change in energy supplies. This will be caused by one or more of these: supply shortages, climate change, demand growth, and political action. To further complicate the analysis, each of these could have one or more reasons. For example, supply shortages could be caused by global peak oil or by major military actions in the Middle East. It does not matter what the changes will be or why they occur, what matters is that there will be major changes in the way the economy works as a result. But because we still do not know the nature of such changes it is difficult to imagine the world of 2050. This is not surprising; if one attempts to put oneself into the initial years of the twentieth century, it would be difficult to see that the coming century would be a century based on oil energy. In 1900, coal was the dominant energy source. Railroads were the backbone of the land transportation system. A US coal miner's strike had the capability to shut down the US economy until after World War II, and in the UK the demise of coal became evident only with the failed miners' strike in the 1980s.

6.1 Changes in Transportation

While the entire economy will change in the twenty-first century, consider the difficulty of seeing what the transportation sector will look like in 2050. As has been noted, this is the part of the economy where any change from oil will have the most profound effects.

Air transport will be the least affected. While there are some early attempts investigating electric options for aircraft, and other attempts to replace fossil fuels with biofuels, the investigations are in their very early stages. Similarly, ocean transport has made some investigations into alternatives to fossil fuels, notably into wind. But neither air nor sea transport is likely to be at the forefront of a new energy economy.

Land transport has more possibilities. This is, in part, because the unit capital investment is so much lower. A truck that carries transcontinental freight costs a fraction of a commercial airplane. Hence, a new technology can be tried with less loss in case of failure. New technologies that become dominant in land transportation will thus mark the transition away from oil.

Most of the developments in moving away from oil for land transport involve using electricity. Options range from purely electric vehicles run from batteries to electric vehicles run from fuel cells to hybrid gasoline/electric power. Improvements in gasoline and diesel efficiency and use of biofuels are being developed to help bridge the transition.

The problem of what comes next is that gasoline and diesel are difficult to replace in the land transportation fleet due to the fact that they provide significant

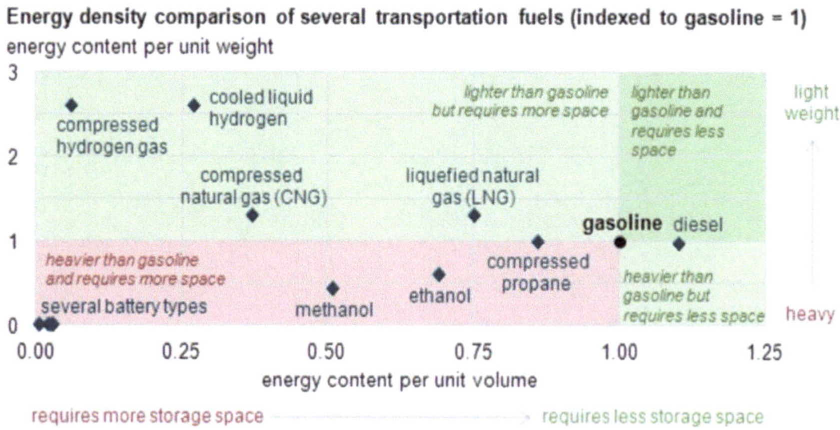

Fig. 6.1 Energy density comparisons (*Source* EIA 2013)

energy in a small space and are relatively lightweight. Figure 6.1 illustrates this advantage.

"Energy density" can be measured with respect to either volume (horizontal axis) or weight (vertical axis). These are physical measures; what will determine their use is their economic cost. All of the alternative options in Fig. 6.1 require more space, which means bigger fuel tanks for the same number of miles; and these bigger tanks will cost more, particularly if they must also be able to withstand high pressure or must keep a gas at very cold temperatures. Nevertheless, the CNG and LNG options in the figure are of particular immediate interest; this is because internal combustion engines running on natural gas are very similar to those that run on gasoline. Indeed, in some cases it is even possible to convert existing engines. Thus if the only issue is oil supply, natural gas options offer an immediate solution. The position of batteries in Fig. 6.1 also explains why electric cars have not yet seen any great success despite their lower operating costs. Figure 6.2 shows an estimate of costs per mile for variously fuelled vehicles.

Figure 6.2 is a simple graphic display of operating costs (cost per mile on the bottom scale) for various electric and gasoline powered vehicles. This varies with the price of electricity or gasoline, and is a function of the efficiency of the particular vehicle. So if gasoline prices are \$3.00 per gallon (US) and the vehicle gets 22 miles per gallon (dashed blue line), the fuel cost will be between \$0.13 and \$0.14 per mile.

The figure shows that the fuel cost per mile driven is generally lower for electric than for gasoline-fueled vehicles. But the difference has not been sufficient to create a mass market for electric cars as yet. There are lots of reasons: the generally higher capital cost of electric cars, "range anxiety" of running out of charge, refueling speed, and lack of familiarity. From the vantage of 2040 we may look back and marvel at the speed of transition away from the gasoline powered, personal automobile transportation (not necessarily to electric cars). But from the vantage of

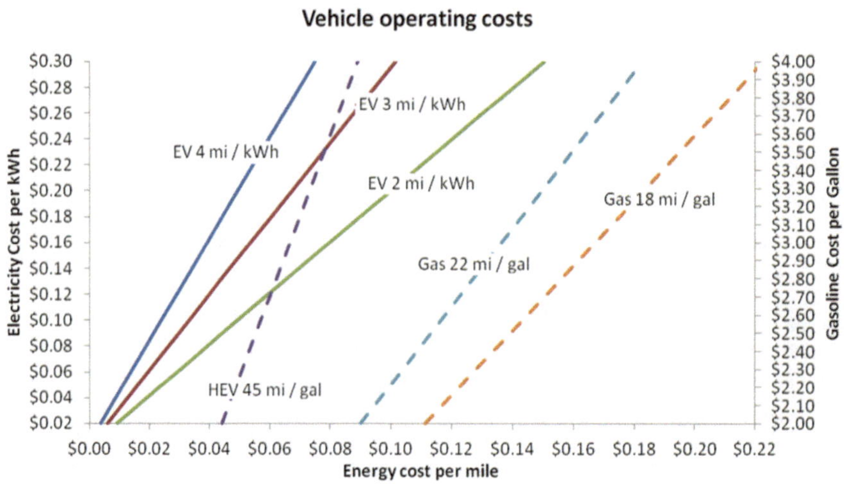

Fig. 6.2 Fuel cost comparison for electric vehicles (EV), gasoline vehicles (Gas), and Hybrid electric vehicles (HEV). *Source* US Dept. of Energy (2011)

2015 market-place economics, the overall most economic alternative for most transportation remains the gasoline or diesel powered private car.

The case for trucks is even more pronounced. Basically, trucks require more energy than cars. All of the disadvantages for non-gasoline or non-diesel cars are magnified in light of these higher energy demands. Sufficiently large batteries are both too heavy and too expensive at present. Range anxiety is not about a weekend trip, but about being able to provide a service and thus make a living. Refueling speed becomes unacceptable.

In short, despite lower operating costs for some alternatives in some vehicles, the overall economic environment for non-oil-based land transportation fuels is not yet generally favorable for change. But, particularly with respect to electric and semi-automated cars, some future changes may be visible on the horizon.

6.2 Additional Financial Disincentives to Transition

As we saw in our economic model, capital is depreciated over the life of the capital asset; basically, with straight-line depreciation, an investment of $1 million for something that should have a useable life of 20 years is considered to be an expense of $50 thousand in each year. To replace this asset after, say, 10 years is to take an expense of $500 thousand in the year that the equipment is abandoned. The early replacement is presumably because something less expensive is available. But note that it has to be **very** much less expensive if the company is to make an immediate or rapid change; otherwise the company will be better off financially to wait and continue to use the old equipment. So most changes occur when capital equipment is being replaced because it has come to the end of its useful life; a different decision might be made today, but the decision was made 10 years ago, and the replacement decision will be made 10 years in the future. This means that the economic advantage of a change must be that much greater if the pace of change is to be forced. Individuals do not do the same formal analyses, but collectively will make similar decisions. For example, Hirsch et al. (2005) noted that cars in the USA are, on average, used for 17 years and therefore it would take at least a decade for cars with significantly better fuel economy to have a major impact on the total amount of gasoline consumed.

Another reason for long time delays in making a change to a new energy economy is the nature of human psychology. We see the world as it is, not the way it will be. In 1905 the airplane had only just taken its first flight and the automobiles were handcrafted. A person thinking about traveling from New York to Chicago would think about an overnight train with a sleeping compartment; they would have had to use lots of imagination to see Interstate highways or jumbo jets. While these options that we take for granted might have been thought of, they were the science fiction of that era. And there were many other options which were just as hypothetical. A major development in the way energy use in transportation evolved was probably something that was not directly involved with transportation at all—the deployment of the assembly line manufacturing method by Henry Ford. This brought the cost of the Model T car down sufficiently to create a mass market, which in turn spurred the development of the road network. Air travel did not really become a major transportation mode until after World War II, when military development of the jet engine changed the energy costs of air transport. Recent news is about self-driving (Ozimek 2014) and electric cars (Geuss 2014). These are improvements to today's transportation structure, but they are improvements in the context of the current framework. It is too early to know whether they will be part of the path to the future or dead ends.

6.3 The Global Energy Transition Situation

The greatest contribution to the global energy picture over the coming decades according to traditional wisdom[2] will be the impact of China, India, and other SE Asian economies. As shown in Fig. 5.4, the energy intensity of the OECD countries has been improving significantly (i.e., the amount of energy per unit of GDP has been going down); were it not for the increase in energy demand from Asia, there would be far fewer concerns about future energy supply. As mentioned, some of the Asian growth, and some of the improvement in energy intensity in OECD countries is due to OECD export of pollution. In other words, energy intensive industries have often relocated from OECD countries to non-OECD countries but the output

[2]See the forecasts referenced in the previous chapter.

of these energy intensive industries is still sold in the OECD (Reinvang and Peters 2008). As Wilson et al. (2002) show, such shifts can have geopolitical consequences. For the situations studied by Wilson et al., the suggestion is that export of pollution makes international agreement on environmental issues more difficult. This in turn reflects back on international trade agreements. In 2014 the Indian government announced that they would not be part of an international agreement on CO_2 reduction because they believe that economic growth and elimination of poverty are more important national priorities (Davenport 2014). Subsequently the Indian government has agreed to join in the 2015 Paris agreement on climate change, but the tensions between limiting emissions and economic growth in all non-OECD countries continue to exist.

More important is that the goal of many non-OECD countries is to have economic standards of living that are comparable to those in the OECD countries. Combined with populations that are growing more rapidly than the OECD countries, the result is significant growth in projected energy demand in the non-OECD economies. For example, the projections of energy use for India include significant increase in personal car ownership and driving due to an expanding middle class, and a consistent high increase in the amount of freight being transported by trucks (EIA 2016a).

China knows it has a major pollution problem (Wong 2014). Two decades ago the skies of Beijing were blue; today the sky in Beijing is most often obscured by the smog. Shanghai controls pollution by rationing automobile registrations (Business Week 2013). There are a host of additional problems in China that make its projected growth problematic. The one-child policy has curtailed the supply of inexpensive labor that was the backbone of China's explosive growth over the past 30 years; the financial system has evolved from a communist philosophy which leaves it subject to new problems such as risk-based regulation of the banking system; and the past 30 years of growth have raised expectations for the future which will be difficult to meet. The top-down nature of Chinese overall planning allows major projects to go forward much more quickly than in OECD countries; the Chinese high-speed rail network is an example. But contrary to the outside perception of top-down control, China's regional governments and nominally state-owned industries are known to simply ignore national directives and policy. A directed campaign against pollution is being implemented. But how much impact this will have remains to be seen due to the imperfections in implementation.

Furthermore, not all such national plans work out very well. One approach is to build new, pollution-free cities. But Chinese jobs are in the existing, dirty cities, so while China continues to build such new cities, there are few improvements in pollution when people do not want to live in the new cities. And the plans themselves require energy to implement, which is also a problem.

Rapid development is prone to bottlenecks that limit growth. Tu (2013) notes that China's rail network is already at capacity with respect to moving coal, and that development of new coal deposits is further constrained by water availability. Hence China is also looking to import coal into its industrial northeastern region rather than mine coal in western China and ship it east by rail.

In addition to developing its own resources, China has been aggressively purchasing foreign oil and gas properties (EIA 2014a). The large number of countries with long term Chinese contracts provides some protection from problems; for example, China has a significant stake in both the Sudan and in South Sudan, although which country controls at least some of these oil fields remains in dispute. China also has agreements with a number of the Gulf oil states. The Chinese have invested in Canadian tar sands, but that investment remains somewhat problematic because it may not be possible to transport the resulting oil out of northeastern Alberta economically. Siberia is a significant future source of oil and natural gas for China, due in part to its geographic proximity. Siberian oil and gas sales to China will also meet Russian aspirations, which has geographical challenges in selling its extensive oil and gas resources. Pipelines that connect to the Russian pipeline network come from Kazakhstan and directly from Siberia. There are also pipelines for both oil and natural gas from Myanmar; while the deposits in Myanmar itself are not large, transshipment of Middle East and African cargos provides both a savings in shipping distance, and more importantly, the ability to bypass the crowded maritime straits of SE Asia.

Whether or for how long all these investments will provide secure supplies in a global energy evolution is open to question. As EROIs go down there will be increasing pressure for new supplies. One needs only look at the recent tensions in the South China Sea (BBC 2014) to see that China remains concerned about its energy supply.

India's growth projections face just as many problems, although there are major differences in underlying reasons from those of China. India's population will increase at a faster rate than China's, which by itself will increase demand. The major issue in India is energy supply, which is currently provided to a great extent by state-owned companies. Most of the fossil fuels are imported, not so much due to insufficient local reserves as to difficulty in converting local reserves to production (EIA 2014b). Prices have traditionally been controlled, and the bulk of indigenous fossil fuel production remains in the hands of state companies. Most of the imports come a short distance from the Middle East to the northwest coast of India.

Despite the best efforts of strategic planners, the projections for both China and India are based on their respective economies of today and existing trade patterns, which means a considerable amount of oil will be needed for their respective transportation sectors. The world oil and gas industry operates on a global basis, and if India and China are still relying on fossil fuels to the extent projected, it is difficult to envision the OECD economies operating on a different basis.

The difficulty of "going it alone" is illustrated by Germany's *Energiewende*.[3] Germany made a conscious decision to shift to renewable energy for as much of its electricity as possible. For the past decade major replacement of fossil fuels was achieved in the electricity sector. While the population has generally been

[3]Literally "energy transformation".

supportive and willing to pay higher electricity charges, patience with such higher costs is beginning to lessen in the face of low oil (i.e., energy) prices. This illustrates some of the difficulties of planning a lower CO_2 energy economy that is different from the global model (Michel 2014).

In the OECD countries, and to a lesser extent in other nations, there is great resistance to moving away from the capitalistic investment model outlined in Chap. 2. Some of this resistance is due to the fact that energy projects have capital investment time frames that are longer than what most people generally consider in their personal planning. As we have seen, moving away from the original investment model may have real financial costs. Although perhaps waning, *Energiewende* continues to have the support of the German population. As one citizen is quoted "There are huge disadvantages in doing it alone, the disadvantages are greater if nobody does it" (Irfan 2014).

6.4 The Global Economic Transition

Some aspects of today's economy will change slowly while other aspects will change more rapidly. The shape of the new economy will be determined by the way in which transportation develops; will we concentrate our populations in high-density cities, or spread out more evenly into low density distributed environments. Each has some advantages. Each would have transportation implications. And the changes in transportation would cascade through the rest of the economy. To some extent these sorts of uncertainty are addressed by using scenario planning (IEA 2013; Royal Dutch Shell 2009), but within any scenario there still remains an increasing range of outcomes going into the future. The past decade has caused considerable change in the energy markets, upsetting patterns of the past 70 years. Thus the range of forecasts will fan out more widely, which means that there are greater uncertainties for those investing.

Tainter (1988, 2010) has noted that societies tend to become increasingly complex. His basic theory is that because there are always problems that need solving, and that solutions to problems add complexity, this progressive increase in complexity is inevitable. But, Tainter observes, the increased complexity is an internal friction; like friction it dissipates energy. Hence maintaining the additional complexity in a society requires more resources—notably more energy. The conundrum is that one of the major problems that the world faces today, exacerbated by the growth projected for China and India, is the question of energy supply. Should this supply continue to be from fossil fuels? Recalling the discussion about peak oil in Chap. 4 there are questions as to whether sufficient oil is even available to support such additional complexity.

Chapter 7
Political Issues

Given the importance of oil in the economy, political issues are bound to arise. These vary from taxation to operating policy to planning for the future. But the role of the oil industry is different in each country, so the nature of these political interactions varies enormously. The issues range from long-term national policies to detailed operational procedures; from very specific taxation issues to major direct influence in the national economy.

The major oil companies, some of which are larger in economic terms than many sovereign states, may be operating in dozens of political jurisdictions simultaneously. The companies are not passive participants; depending on the situation in each country, the companies may influence or even dictate government policies. In other situations, a government may use oil companies within their jurisdiction as a means of implementing policy.

7.1 Taxes and Subsidies

Because few economists understand the economy as an energy flow, as discussed in Chap. 5, even fewer politicians look at the energy business in this way. There are a huge number of ways in which oil can be taxed or subsidized, and countries frequently do both. As we saw in Chap. 2, the financial prospects of a particular project can be heavily influenced by taxes; it is not only the tax rate that matters, it is also the accounting rules about what can be considered profit. And it is not always clear whether a country's policies are encouraging or discouraging to fossil fuels. Norway, for example, has comparatively high taxes on both production of oil and on retail sales. At the same time, Norway's tax policies allow companies that are doing exploration in one area to offset many of their expenses against revenue from a completely different area. The result of this tax offset is that companies that have Norwegian production are, in effect, having their exploration activity subsidized by the Norwegian government in the form of reduced tax payments. This is a

S.W. Carmalt, *The Economics of Oil*, SpringerBriefs in Energy,
DOI 10.1007/978-3-319-47819-7_7

significant encouragement for Norwegian exploration activity as the direct result of tax policy. At an informal meeting, I had the pleasure of hearing two economists argue as to whether Norway was subsidizing or heavily taxing petroleum.

Another example is in the USA where the tax code has allowed a "depletion allowance" for produced oil and natural gas. This is justified as being akin to the depreciation that is allowed against physical capital assets like the well in our analysis from Chap. 2; it is a non-cash expense in a company's accounts which offsets the value of the reserves. By allowing a company to show such an expense, the company's income, and therefore income tax, is lowered. Industry groups maintain that this is not a subsidy because the government does not pay anyone; but the economic effect is very similar. The government clearly has foregone some revenue and the company has received a benefit. In the USA, the depletion allowance remains a part of the tax code, although with many detailed rules designed to limit its applicability so that the largest companies do not benefit (IRS publications 535 (2015)).

Norway and the USA are stable democracies, so changes in taxes are made peacefully. In other countries, particularly those which subsidize retail fuel sales, changes in a tax or subsidy regime may be more difficult. Venezuela is (or traditionally has been) a significant oil-producing country and has sold gasoline and diesel in the local economy for far below its cost (Wagner 2013). Increasing the price of fuel so that it more realistically relates to either production costs or the price of other goods and services frequently cannot be done because populations object (Los Angeles Times 2016). In such a situation, the deferred underlying economic problems make the political situation less stable. While the situation in Venezuela is extreme, a number of other countries also sell gasoline and diesel at prices that are below cost (Wagner op.cit.).

But the stable democracies have their problems as well. As has been discussed, energy projects in general, and oil and natural gas production in particular, are long-term, capital-intensive projects. Once a project is underway, a change in taxes or other assumptions of the financial model can convert what was a profitable undertaking to an unprofitable one. Yet it is a delicate political balance to provide a phase-in period for a new policy without causing undue financial hardship by changing the rules abruptly.

7.2 Market Stability

Two historical truths about the oil industry will continue to have a significant impact on the industry for at least several more decades.[1] The first is that the industry has almost always had prices controlled by some sort of oligopoly or

[1]See Sampson (1975) and Yergin (1991) amongst many others.

cartel. Rockefeller's Standard Oil Trust was followed by the Achnacarry Castle agreement which in turn became the 'seven sister'[2] group outside the USA while the Texas Railroad Commission took care of maintaining a reasonable price (from the perspective of the oil companies) within the USA. These arrangements have had at least tacit support of governments. The oil companies, based in the USA and Europe, have benefitted by being protected from cut-throat competition. But as global production shifted to other countries in the 1960s, many of which had poor, indigenous populations that would benefit from higher crude oil prices, the major producing countries founded OPEC, the Organization of Petroleum Exporting Countries. During the 1970s, OPEC wrested the monopoly control away from the seven sisters, frequently by forming national oil companies which the government then required to have a major role in all activities within the country. Saudi Arabia, which is by far the largest producer in OPEC, has been able to maintain market order by being the "swing producer," with the government converting Aramco into a national oil company.

The second truth is that the Iran-Iraq-Saudi Arabia-UAE area is an especially "oily" part of the world. The first discoveries in this region were made in Iran in 1908[3] and in the period immediately following World War II the Gulf[4] area became the predominate oil producing area of the world. Table 7.1 shows this importance.

For reasons that are beyond the scope of this book, the Gulf region has been, and remains, one of tenuous political stability. Iraqi exploration and production has been limited ever since the start of the Iran–Iraq war in 1980. There is much industry hope for new discoveries and enhanced production given the technology improvements over the past 35 years, but the country remains politically precarious (IEA 2012). Iran's contribution to global production has been limited by sanctions which are the indirect result of a different Middle East political problem, the Israel-Arab impasse.

The above percentages illustrate the critical role that Saudi Arabia plays in both the Gulf region and in the global petroleum industry. In September 1985, Saudi Arabia cut the price of its crude oil almost in half; this is reflected in the drop of average oil prices from 1985 to 1986 from $27.53 to $13.30 per barrel. While this move established Saudi Arabia's control over both OPEC and the global oil markets, this role of "swing" producer was established more than one-quarter century ago. Several factors, all of which are hotly debated, question whether and under what conditions this role can be maintained.

---

[2]The seven: Exxon (Standard of New Jersey), Mobil (Standard of New York), Standard of California (Chevron), Texaco, Gulf, BP and Shell. Today there are but four: Exxon and Mobil have merged into a single company and Chevron has absorbed Texaco and Gulf.

[3]See the histories already mentioned.

[4]The Iranians are not Arabs, and the Arabs are not Iranians; each feels strongly about this. Thus 'Persian Gulf' and 'Arabian Gulf' are both unacceptable names for this body of water. So it is simply "the Gulf." But note, this Gulf is not to be confused with the Gulf of Mexico, which is the center of the US oil industry and which is also called 'the Gulf' when discussing US issues.

Table 7.1 Saudi Arabia production in millions of barrels per day with comparison to other sources

Country/Year	1950	1970	1990	2010	2015
Saudi Arabia	750	3851	7105	10,075	12,014
As a % of OPEC	*18.8*	*16.9*	*29.8*	*28.7*	*31.4*
As a % of world	*6.9*	*8.0*	*10.9*	*12.1*	*13.1*
All Gulf Countries[a]	2133	13,741	16,900	24,893	29,812
As a % of OPEC	*53.3*	*60.4*	*70.8*	*70.9*	*78.0*
As a % of world	*19.7*	*28.6*	*25.9*	*29.9*	*32.5*
Total OPEC	4000	22,762	23,857	35,088	38,226
As a % of world	*36.9*	*47.4*	*36.5*	*42.1*	*41.7*
Total World	10,852	48,056	65,385	83,296	91,670

[a]Saudi Arabia, Iran, Iraq, Kuwait, Bahrain, Qatar, UAE, and Oman (data primarily from BP statistical reports (2014, 2016); 1950 estimates from multiple sources)

In the mid-1980s Saudi Arabia still enjoyed the luxury of having a relatively small population; in addition, it had built up massive foreign currency reserves during the decade of the 1970s. As a result, it was in a position to reduce production and forego income. Other OPEC members, some with very much larger populations, could not afford any cutbacks because they needed every dollar of petroleum revenue. On the other side, the very large and easily accessible reserves in Saudi Arabia made it possible for Saudi Aramco, the Saudi national oil company, to increase production. Today the situation has changed. Saudi population has doubled from 13.3 million in 1985 to 27.3 million in 2001 (UNDESA 2012). Furthermore, this population has become used to the income provided from oil revenues. Hence the ability of Saudi Arabia to absorb significant income reduction over any extended period must be questioned. On the question of increasing production to cover shortages in the market and prevent major price increases, the Saudi government has announced that they can produce at a rate of 12.5 million barrels per day if needed. But the ability to maintain such an increase over any extended period of time is questionable (Said 2013; Maschhoff 2013).

7.3 Geographical Issues

The fact that almost 30% of the world's oil, and hence a significant part of the energy for the world economy, comes from the Gulf countries makes the geopolitical situation in the region of considerable economic importance. That the Saudi government, which is to say the Saudi royal family, is particularly targeted by both Sunni fundamentalists (Baer 2003; Crooke 2014a, b) and Shi'ites (Wehrey and Sadjadpour 2014) is of particular note. Furthermore, geopolitical tensions are not limited to the Gulf. Russia, which is a significant oil and natural gas exporter, has been subjected to sanctions for its actions in Ukraine. This continues to have the potential to disrupt supplies in Europe, which has been Russia's main purchaser of oil and natural gas.

The volume of oil that flows through the production, transportation, refining, and distribution system is vast. Throughout the system there is storage to facilitate day-to-day operations; the amount stored in each automobile's gasoline tank at any given time is an example. But volume makes storage of any significant reserves for emergencies impractical; individuals do not store gasoline sufficient for the entire next month, let alone the entire next year. Major long-term storage faces similar problems. The only practicable way to adjust output to demand is to increase or decrease the amount being produced at the source, using the original geology as the storage.

The consequence of this is that sudden shifts from one producing area to another, even if the absolute amounts are available, are not feasible. The entire transport infrastructure is designed for the current geographic distributions. An example of how difficult this can be is that production from the Bakken areas of the USA is being constrained by lack of transportation between North Dakota and refineries (Craig 2014; Hussain 2013).

It is in this context that what might be called a "special relationship" between Saudi Arabia and the USA needs to be evaluated. Basically, the US government, either openly or quietly, has committed itself to defend the Saudi government ever since oil was first discovered. This was visibly confirmed in the aftermath of Iraq's temporary takeover of Kuwait in 1990. In return, the Saudi government has advanced various US interests over the years; at present, there is speculation that there may be a current understanding in which the US provides defense assurance to the Saudi government and in return they use their oil marketing power to depress global oil prices, thereby increasing the effect of sanctions on Russia (Engdahl 2014). But this can only be speculation, because the specifics of US-Saudi understandings are confidential.

7.4 Climate Change

The political giant in the room is the issue of climate change. The CO_2 emissions issue has been recognized for decades but as yet there has been no concerted political action to mitigate damage. One of the problems with a government taking independent action with respect to climate is the tragedy of the commons problem. This basically holds that an individual's economic decisions may be uneconomic for a shared resource. The classic example is overgrazing of common pasture land used by all the farmers in a community. The pasture will provide well-fed livestock for everyone, provided it is not overgrazed. The benefit to an individual farmer of adding one extra animal to his herd or flock is greater than the decrease in overall benefit that this addition causes to the group. If every farmer does this, the common pasture becomes increasingly overgrazed and everyone loses. In addition to common pastures, fisheries have the same problem. In a classic article on the subject, ecologist Garrett Hardin (Hardin 1968) observed that this is not a problem that can be fixed by technology, but requires a social change. In future decades the disposal of CO_2 into the atmosphere may be cited as an example of this economic problem. It is almost certainly the reason that the global negotiations to limit CO_2 emissions made no progress for over 20 years, despite the increasing urgency of the problem.

There is a case to be made that many present uses of fossil fuel are so inefficient that there is an economic benefit to be gained simply by increasing the efficiency of their use (see Irfan 2014). Adopted widely, such efficiency improvements can have a significant impact on global CO_2 emissions. Because they are economically beneficial without respect to the global atmospheric commons, such changes can be implemented locally. National, or even local, government encouragement of such efficiencies are likely, meaning that the specific taxes and charges used in finance decisions are likely to change. Such changes will provide short-term benefits to both the implementers and to the global beneficiaries of the atmosphere, although it defers rather than solves the overall CO_2 emission into the commons problem. But to insist on only global solutions to CO_2 emissions is to make the perfect the enemy of the good.

Nordhaus (2014) has proposed a practicable approach to the international impasse on climate negotiations. He suggests that countries form "climate clubs." To be a member of the club, a country will need to meet specific criteria with respect to CO_2 emissions. But the benefits of being a club member, specifically tariff differentials, compensate for whatever costs are incurred. Basically, the tragedy of the commons arises from the benefits not being commensurate with the costs due to the benefits being shared globally and the costs being borne locally. Nordhaus, whose research uses complex global economic models, suggests that a club which taxes CO_2 at between \$25 and \$50 per ton would attract most countries as members. In essence, he suggests that if only some users of the commons, a "club," adopt rules that prevent "over grazing" they can effectively enforce these rules on others.

Political pressure is growing to address the issue of climate change, but it is still unclear how this will be done. The "climate club" is but one proposal.

The Carbon Tracker project (Leaton 2012), based in London, raises another political worry. It notes that the publically traded oil companies put a monetary value on oil reserves still in the ground. While these numbers are conservative with respect to the amount of oil that may be technically available, they are actually very generous when viewed from a CO_2 emissions perspective. Between 70 and 80% of the world's already-discovered oil reserves simply cannot be produced (or more accurately burned) if climate change goals are to be met. The problem has been given a name: "stranded assets" are financial assets that cannot be used. There are important financial implications. Reducing the financial assets of companies to reflect that some of these may be stranded will, in principle, lower the value of the company's worth and thus its share value. There is speculation that full recognition of this stranded assets problem could result in a major sale of oil company shares. Financial markets move much faster than changes in political terms, which in turn move much faster than changes in underlying technology and energy use. If the financial and investment community becomes concerned about such stranded assets there is the possibly of a financial crisis (Carrington 2013).

There will always be losers when economic policy changes. Governments can effectively change asset valuations or destroy profit forecasts with the stroke of a pen. Compensation for lost asset value is not often discussed at all (Hayes 2014). The political dialogue around profit forecasts tends towards blocking changes to currently profitable activity as opposed to allowing existing project financing to be gradually incorporated into the new policy environment. The objections may be philosophical (Oreskes and Conway 2010) or grounded in vested financial interest (Manne 2012).

The primary climate change concern is CO_2 emissions. As weather events increasingly illustrate climate change it is likely that the political pressure will grow to address the CO_2 and climate change concerns (Lifton 2014). Indeed, the major oil companies all presume this and already include it in their economic forecasting and decision-making (Schapiro and Scorse 2014). The net result of the political environment with respect to climate change is that there may be major changes in the political willingness to "do something" by 2020. While this is a strong possibility, there remains great uncertainty as to what form such changes might take. This uncertainty makes financial models less accurate, and thus inhibits investment. This is particularly true for the oil industry, where most projects have especially long lives, meaning that planners have to make assumptions not just for the next decade, but for the next half-century.

The critical point to realize is that the technical costs and project financing, together with the geological availability of fossil fuels, are but a part of the future of the oil and gas industry. The "hard science" of these aspects of the business is implemented in a political setting that is always changing. This political setting has a major impact on the evolution of the business.

Chapter 8
Forecasting Natural Gas and Oil Production and Use

The oil and natural gas industry is rapidly evolving. Because these two fossil fuels are so closely aligned in geologic occurrence, production technology and final use they need to be taken together. How are they likely to evolve over the coming decade or two?

Reviewing the oil situation from Chap. 4, the oil production peak for "conventional" crude oil deposits has passed. While there is plenty of oil still to be had in the ground, the EROI for new discoveries has decreased significantly and will not return to old values. All of the new prospects are what the industry calls "high cost oil." There are a number of types of such high cost oil: Arctic areas, deep-water prospects, tar sands, and onshore tight oil resources. Even the largest companies have capital constraints, and these will direct which of the expensive oil opportunities will be exploited first.

These resources have already been discussed in previous chapters. The reviews below focus on the risks they have, and thus the likelihood that they will or will not become part of the world's energy future.

8.1 Harsh New Environments

Both Arctic areas and deep water exploration programs are seeking traditional oil deposits; that is, deposits where oil that has been trapped after migration into reasonably permeable reservoirs. The geology and production are similar to oil fields that have been developed over the past 100 years. As such, they are a known quantity. The challenge and expense result from the harsh environmental conditions, which require capital investments in the hundreds of millions and frequently billions of dollars into projects that often only have initial financial returns after five or more years. Cash break-even times can easily approach a decade, despite projections showing profitability overall extending over a longer period. These high, up-front capital requirements for individual projects make it difficult for even the

S.W. Carmalt, *The Economics of Oil*, SpringerBriefs in Energy,
DOI 10.1007/978-3-319-47819-7_8

largest companies to avoid gamblers ruin, even when risk is spread through partnerships, joint ventures, and other risk-mitigating techniques.

The development of the Kashagan field in Kazakhstan offers examples of these types of problems (Scheck 2013) and thus the risks they entail. The field is a giant, with reserves estimated in the 10 billion barrel range. Discovered in 2000, production has been pushed back again and again due to various problems. The field is located in the Caspean Sea, but the water is shallow. In fact, it is so shallow that the development of the field requires a combination of offshore and onshore technologies, technologies that have frequently had to be engineered specifically for this field, thus making development just as expensive as a deep-water environment. The location is remote, requiring materials and people to be moved and supported. The climate is harsh, with winter ice causing unforeseen problems in the shallow water. And the list goes on. Frontier areas in the Arctic and deep water oceans have similar challenges and resulting expenses in new and unfamiliar conditions. Remembering the importance of timing in project financing, it is easy to see why the major oil company ownership interests in the Kashagan project have repeatedly changed over the years.

8.2 Tar Sands

The tar sands, while also requiring massive capital investment, have different risks. For tar sands, both the geology and the extraction technology are reasonably well-known. Hence the amount of recoverable oil is known and the production risks can be managed. Some of the Canadian deposits are mined. The bitumen must be separated from the sand and subsequently processed into something that approximates crude oil.[1] But much of the known Canadian occurrence, and all of the Venezuelan occurrences, are too deep to be economically mined. The bitumen must be extracted by a combination of in-situ heat and chemical processing. It is only after being converted into synthetic crude that the produced oil can enter the market for crude oils. While all high-cost oil is subject to market risk, the high preliminary processing costs make high oil prices imperative for these projects. Thus oil price forecasts are the primary risk factor.

In conventional oil deposits, the high cost is up-front capital expenditure; operating costs are not the major expense in the initial economic analysis. In economic terms, once in production, a deep-water oil discovery has relatively low marginal production costs. This means that most producing oil fields can continue to be produced with a current operating profit, even when prices fall. While falling oil prices may make the project less financially attractive in total, once the front-end

[1]This product is one of several different types of synthetic crude oil, with one of the first Canadian companies to exploit the tar sands named Syncrude. Other types of synthetic crude can be made, most usually from coal.

capital has been spent, a lower oil price will not have so much impact on production rates and volumes. There will be less of a margin to repay capital, that is to meet interest payments and pay dividends, so companies that have only high-cost oil in their portfolio will be squeezed financially. But looking forward the physical operation is likely to continue rather than be shut down. Indeed, lower prices will tend to encourage increasing production in order to maintain revenue forecasts.

Not so with tar sands. While there are major capital costs in opening a mine or setting up an in-situ extraction scheme, there are also significant ongoing operational costs in converting the resource into synthetic crude. These higher operational costs mean that a lower global oil price can fairly quickly make even the ongoing operations unprofitable.

8.3 Tight Oil

When we turn to tight oil, we find still another different situation. The amount of capital involved to participate in a tight oil play is much lower, and it can be spread over a larger number of independent risks. Industry advocates of tight oil resources speak of "factory drilling" or "assembly line drilling". This approach addresses two problems simultaneously; first, it lowers the per-well cost, and second, the revenue from the first wells drilled is available to pay the costs of drilling the additional wells.

The risks involved in any given well in a "tight" oil play are somewhat different than in a conventional play. In a conventional play it is necessary to find a geologic combination of source, migration into a porous reservoir, and finally, some trapping mechanism. Once the geological source deposits have been identified in a basin, much of the risk in conventional plays comes from problems with the other factors—most often either the migration into the porous reservoir rock has not occurred as predicted, or the trap is not effective and the migrating oil has passed through the trap and been lost over geologic ages. In "tight" plays there is little scope for the oil to migrate, and after the initial wells the province is known to have the required source and the oil or gas is still there. Thus, while there are many "dry holes" in conventional oil exploration, there are far fewer in tight formations. Rather, the risk changes to one of extraction rates. The details of how best to gather the oil out of the impermeable layers into the horizontal well are primarily engineering issues. There is the risk of not "landing" the well in the best horizontal layer, the risk that the fracking of a particular stage will not go as planned, and the risk that the detailed flow into the well will not be as predicted. The rapid production decline requires a large number of wells. The result is that planning a well-drilling program over a period of time may include dozens, if not hundreds, of wells. Remembering our financial model again, the rapid decline of production rates in "tight" plays makes the initial few months of production revenue critical to the financial health of the overall operation. In this sense, the financial model begins to be similar to that of a factory, where there are some initial costs for construction (exploration concept and land acquisition), after which there develops a steady state production situation, with costs for raw materials (the cost of new wells) being paid for by the oil sales from all previous wells. Adoption of such "factory drilling" approaches is reported to reduce costs by up to 40% (Forbes and Wilczynski 2010). Even so, if there is a marked decrease in revenue from new wells, the entire program is put in jeopardy. Some analysts have serious doubts that the level of continued production can support such a continuing operation over any length of time (Patterson 2014), and others suggest that the way by which high initial production rates pay for additional wells is nothing more than a Ponzi scheme (Washington's Blog 2014). In addition, while the area of a tight oil "play" may be large, there must be boundaries, so this process cannot continue indefinitely. For all these reasons there are indications that the production per well for tight oil, at least in the Bakken play, has already peaked, with the best horizons having been identified and exploited. (Brandt et al. 2015).

The combined use of horizontal drilling and fracking to exploit tight reservoirs has only been widely deployed for about a decade. This implies that the industry has considerable room in which to make technical improvements on these techniques. For example, exploitation of a tight reservoir is typically done from a "pad" rather than each well having its own surface installation. Drilling rigs that can steer a well horizontally are much more complicated than rigs that just drill straight down, and therefore more expensive to operate. So a company will start with a simple vertical rig on a pad and drill all the projected wells from that pad down to near the target level (Frantz 2014). The more expensive, horizontally capable rig will then be moved in and will drill the horizontal portion of the well. Depending on the specific play, this will save several days of expensive rig time for each of the pad's wells. Another savings is to make the pads somewhat larger so that more wells can use the same infrastructure. The resulting savings add up, as can be seen in the decline of the per well costs over a two-year period in the Bakken play, as shown in Fig. 8.1.

Because developing tight oil and gas via the factory drilling model involves primarily the expense of on-shore wells, a company can more easily adjust the tempo of its activities according to the current price of oil or gas. In the current environment of low oil prices, some companies that were financially very highly leveraged have had problems with reduced activity, but others have been able to weather the past year of very low oil and gas prices surprisingly well (Durham 2016). The jury is effectively still out as to whether it will be profitable overall for companies to engage in the USA's tight oil production increase. There are serious critics (e.g., Hughes 2014; Berman 2015), but the major companies that continue with this activity are indicative of internal financial models that project profitability.

Two points already made are worth repeating with respect to tight oil and gas: first, that being profitable is not the same thing as providing the energy that the economy will need, and second, that some companies can be profitable in specific situations while the industry activity, as a whole, is not.

There is also considerable question as to whether the tight oil technology is applicable to areas outside of the USA. There are certainly geological areas that are known to be attractive prospects for these techniques on a technical basis. The major difference between the USA and other locations is the private ownership of

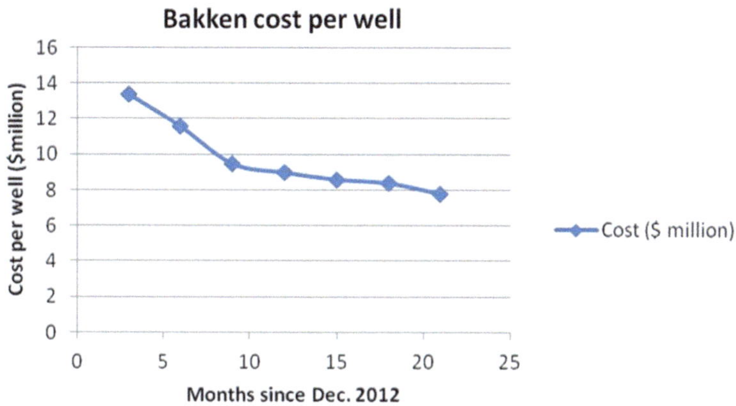

Fig. 8.1 Costs per well in the Bakken play (*source* Pan 2013)

the oil. As already mentioned, the law in the USA is that the owner of the land surface also owns the oil thousands of meters below it. Historically this has meant that the oil industry in the USA has had the encouragement of individual property owners to drill wells—if the company finds oil or gas, the landowner will receive a royalty. The payment to a landowner for any oil found is generally between 12% and 25% of the oil or gas when produced. Outside the USA (and a few other places), it is the government that typically owns resources at depth. Thus, the local landowners do not benefit much; indeed, they may object to the surface disruptions and environmental risks, becoming a political force opposing such operations. Furthermore, in many jurisdictions the government is more likely to expect at least 50%, and often more, of any proceeds. The financial prospects for a well need to be significantly better in such cases than in the USA where the tight oil and gas activity has been concentrated to date. Not only does this significantly change the economic prospects for a company, but it also means that the structure of the industry is somewhat different. The USA has many small independent oil companies, supported by a host of service contractors, drilling many wells, whereas outside of North America the availability of rigs, service contractors and other support is more limited. This can be seen in the recent number of drilling rigs in 2014, with 1927 rigs in the USA and 1749 in all of the rest of the world (Baker Hughes 2014)[2]. Almost one-quarter of the non-USA rigs are in Canada.

As discussed in Chap. 4, there is debate about whether the peak of "peak oil" has been reached. Presuming it was reached in 2005 for conventional oil, this still this means that perhaps there is approximately as much conventional oil to produce in the future as has already been produced. This additional oil will be partly oil that has already been discovered, and partly will be new discoveries of conventional oil.

[2]Baker Hughes, which makes drill bits, has published the number of currently operating drill rigs since 1944. This number is now widely reported and used as an industry statistic.

Of this, much of the oil that has already been discovered has low production costs – it was discovered when oil prices were not so high, so the investments were made in that fiscal environment. But traditional economic theory holds that the market price will be determined by the marginal cost of the last unit. Hence the expensive new oil will result in very handsome profits for those controlling the reserves already discovered, but not yet produced. Much of this low-cost oil is in Saudi Arabia and its neighboring countries, where political conditions could render it unavailable. Presuming it is produced, it remains to be seen whether the high profits will be used for investment, local development, jihad, or something else.

8.4 Natural Gas

Oil companies are also gas companies, so their futures have similarities. As has been noted, gas has not usually been the objective of hydrocarbon exploration. Nevertheless, gas is frequently found when exploring for oil. It is found both associated with oil, and on its own. In addition to hydrocarbon locations, gas is found in association with coal; indeed, because methane can explode, it is a danger in coal mines. But in coal strata too deep to mine economically, not only can this gas be recovered by drilling, but coal to methane conversion can be done in situ. As a result, there is not much danger that we have reached "peak gas". The IEA estimates that, in energy terms, there is the same order of magnitude of natural gas in declared reserves as there is of oil (190×10^{12} m^3 of gas $\approx 1.2 \times 10^{12}$ oil-equivalent barrels of natural gas versus 1.6×10^{12} barrels of liquid oil) (IEA 2013; BP 2014), with many additional resources already identified.

It is impossible to do a statistical peak gas analysis in the same manner as for peak oil. Over the history of oil production, so much gas has been flared as an uneconomic by-product that the figures for gas upon which to base such a statistical analysis are not available (Deffeyes 2005). And flaring continues, although it is reduced compared to past decades (World Bank 2012). This reduction in flaring is for several reasons; the first is because of an increasing ability to market the gas as fuel; the second is because of CO_2 emission awareness, reinforced in some cases by regulation; and the third is because it is a useful product that can enhance oil production in-situ. Nevertheless, considerable flaring still goes on, which underscores the difficulty of using this resource within the current financial framework. The World Bank's Global Gas Flaring Reduction program estimates that almost 1 billion barrels of oil equivalent is being flared each year (World Bank 2012).

Were it not for climate change it is likely that natural gas would be the primary energy source of the twenty first century. The major developments required are in arranging the financing for the more expensive transportation that natural gas requires on a large scale.

Moving gas from gas wells to users requires either pipelines or specialized, purpose-build shipping. The capital cost of such projects is huge, and once the investment is made, the resulting infrastructure is inflexible. For example, the

Russian "Power of Siberia" pipeline that will run from the Yakutsk gas field in central Siberia to markets in northeastern China has cost estimates from $20 billion to $70 billion (Korchernkin 2014; Hanner 2014). Similarly, the now-abandoned South Stream pipeline to transport gas from Russia's Black Sea coast to Bulgaria had a price tag of $10 billion for just the Black Sea portion (Itar-Tass 2014), with at least twice that amount being required for the various onshore portions of the complete transport route. To finance such projects, some sort of long-term sales contract is needed. Hence the recent Russian–Chinese 30-year contract that is the foundation for the construction of the "Power of Siberia" pipeline (Perlez 2014).

Liquified Natural Gas (LNG) projects are not necessarily cheaper. Cost considerations to construct an LNG "stream" are laid out in a 2007 article by KBR (Kotzot 2007), a major industry construction contractor. LNG is generally shipped by specialized tankers, so the processing plant consists of the facilities needed to accept the gas at the plant entrance, where it has been delivered by pipeline, clean and cool it to a liquid, and store it until it can be loaded onto an LNG tanker. The total cost of such a "stream" is on the order of $1 billion. From well to customer requires adding costs of pipelines to get the gas to the LNG plant, the construction of tankers to support a cryogenic liquid, appropriate port facilities at both the loading and unloading ports, a regasification plant, and the final pipeline distribution to customers. In conjunction with a major new gas discovery, such as the Mamba field in Mozambique, one must also add the full gas field development costs, bringing total cost estimates to about $50 billion (Offshore Technology 2014). If the political risk is too high, the field will not be developed; if the combined costs cannot be supported by long-term contracts, the entire project will be abandoned.

As the globe was explored for oil in the twentieth century, many gas fields were discovered that were simply left undeveloped. The high oil prices of the past decade have regenerated interest in many of these, and plans are being drawn up for exploitation. Two things will happen as a result: first is that natural gas infrastructure will be installed and expanded. For example, LNG projects in the USA already presume a natural gas infrastructure in the Gulf coast and eastern parts of the country. While new pipelines will be needed to move gas from the Marcellus areas of Pennsylvania to a US Gulf coast LNG facility, the existing infrastructure allows this need to be met in several ways: gas that was being shipped from Texas to the northeast will be the source for LNG projects in the Gulf coast area, with northeast customers being served from the Marcellus; some existing pipelines will be expanded; some new ones will be built; some will have connecting stations and pumps installed so that they can be bidirectional, depending on local market needs. In short, there will be many smaller projects which will combine to increase the overall network capacity. In general, the increased infrastructure also will enable more flexible market responses between supply and demand, thus making any long-term fixed-price contracts more risky. And second, considerably more projects will be proposed than will actually be built.

But where such diverse infrastructure is not already developed there are many questions. For example, what will happen to the vast amounts of gas available in southwestern Iran? Some Iranian gas already goes as LNG to East Asia, and additional LNG facilities are under consideration to meet market growth in Asia. But this does not use the entire potential resource.

Both long-distance pipelines and LNG projects have significant operating costs. In the case of LNG the cost of cooling methane to below −164 °C requires significant energy; preparatory processing, and transportation will use additional energy. The most modern LNG streams can expect a 10% loss of the gas coming into the facility just for the cooling (Center for Energy Economics 2006), and the loss may be higher if the gas has impurities that must be cleaned prior to liquification. For pipeline transport, the loss comes from the need to have periodic compressor stations along the length of the pipeline. Its not that the gas is leaking (although some may) but rather that as it moves through the pipe there is internal friction, which dissipates the mechanical energy of high pressure into low grade heat energy. Whether transported by pipeline or LNG the fuel used is frequently the natural gas itself, with the result that more gas needs to be produced than is available for sale to the final customer. As a very rough guide, between 15% and 20% of gas produced at a gas field, at a minimum, will be utilized in getting the gas to a user.

The global statistics for the oil and gas industry are, at best, confusing. Are reserves really increasing? The IEA statistics indicate that not only are the amount of oil and gas being produced each year increasing, but so too have reserves increased, at least until 2015. From Hubbert's insight that because oil must be discovered before it can be produced this reserve growth is, therefore, a forecast of potential production growth. But not only is it difficult to assemble the global figures for the capital investments made to support these data, but some of the data are puzzling—in particular the fact that some countries seem to produce significant amounts of oil without announcing either new discoveries or reserve declines (Aleklett 2012). The major oil companies are showing the opposite: lower reserves despite spending massive amounts in exploration. The combined exploration budgets of ExxonMobile, Shell and Chevron come to over $100 billion per year and they are not replacing the company reserves consumed during the year. As one industry analyst observed: if an oil company does not replace its reserves it eventually goes out of business.

In this context it is worth remembering that John D. Rockefeller built Standard Oil on the basis of a refinery, not on exploring and producing oil. The twentieth century energy outlook was that that having reserves in the ground was the critical control point. It may be that the energy control points in the economy of the twenty first century will be elsewhere—perhaps in control of pipeline networks and/or long-distance electricity transmission, for example.

8.5 The Changing Nature of Energy Markets

The markets monopoly, oligopoly and cartel structure of the oil industry already described means that free markets are a relatively recent phenomenon. Until the 1970s, the important oil price was the "posted price" (Oweiss 2014), which was the nominal price of oil. Initially based on a cargo loaded onto a tanker at a US Gulf port, the posted price was a basis from which were added transportation and quality differentials. In the days of the Seven Sisters, the companies actually owned the oil in the ground, and for the most part shipped it in ships which they operated to refineries that they also owned and finally sold it through distribution channels which they owned. This vertical integration was a way of doing business that went back to the Standard Oil monopoly. There were several reasons for this. First, oil is imperfectly fungible. It can be light or heavy, light being shorter hydrocarbon chains and heavy being longer chains. Some of the heaviest oils are more dense than water, and hence sink rather than float. The oil can also have more or less sulfur, less being called "sweet" and more being called "sour". Refineries are built to process a particular type of crude. So while a refinery may be able to accommodate a crude oil which is somewhat different from what it was designed for, there are limits to how different and how much of the different crude it can process. As oil moved through the vertically integrated oil companies, it was "sold" from the producing subsidiary company to the transportation subsidiary company to the refining subsidiary company and then finally to the consumer. The refineries were built to process the crude oil that the producing company had available. The posted price mechanism allowed the company to make its profit in whichever subsidiary company would be subject to the least tax; hence the posted price allowed the industry to adjust somewhat to supply and demand while maintaining the entire profitable structure. The companies outside the system would buy or sell into this flow of oil at the prices based on those used for the company's internal flows of oil. But these additional sales were not the major part of the oil supply, and thus did not have much effect on the prices at any stage.

The fact that the posted price was kept low for the various producing countries and the increase up to market price was made within the company in ways that maximized the company's tax advantages and profits was the principle reason that OPEC was founded in 1960. When the Arab Oil embargo was imposed in 1973, this pricing system was the system being used. More important than the temporary embargo was that OPEC took over the setting of the posted crude oil price, raising it, approximately, from \$3 to \$12 per barrel. Since the 1970s, the spot markets have become increasingly important in setting oil prices, with both the London and New York financial centers having standard contracts—Brent Crude in the case of London and WTI (West Texas Intermediate) in the case of New York—against which most other contracts are referenced. For oil, then, prices around the world tend to move up and down on a global basis. Both London and New York have futures markets based on the standard spot contracts, and the full range of financial trading, speculation and hedging occurs. As a result, there is confidence that investments in oil can be marketed in this global environment. At the same time, this means that oil companies are no longer able to internally manage their profitability via shifts in internal pricing.

In contrast, gas markets are much more fragmented and uncertain. The high capital costs of installing the somewhat inflexible infrastructure to get the gas to market are added to the up-front capital requirements of developing the field. Together, this requires long-term financing. Long-term financing requires long-term contracts, and these contracts tend to tie producer and seller together. An example is the Russian–Chinese 30-year contract previously mentioned. But what price should such long-term contracts specify?

When natural gas first started being developed, the market for gas was local. Most cities had gas supplies, but generally it was gas made from coal, and whatever the source, the pricing was specific to the local situation. When significant natural gas was first exploited in the Netherlands, the decision was to price it relative to the energy equivalent of oil. Much of the natural gas network in Europe, with pipeline sources today in the North Sea, Russia and North Africa, and with additional amounts of gas imported as LNG, is still constrained by pricing under long-term contracts that are tied to the price of oil. In the USA, natural gas prices were fixed in interstate commerce until the 1980s; initially they were low, because as a major oil producer the USA had significant natural gas available as a by-product of oil production. But the interstate regulated price did not keep up with energy content or oil prices, which led to gas shortages in the interstate market and high prices in strictly intrastate markets, notably Texas. But when the USA price was deregulated, a spot market quickly developed because by this time there was an extensive pipeline network in place which could connect a large number of independent producers and consumers. The natural gas markets of eastern Asia (Japan, Korea, and northeastern China) are almost entirely dependent on LNG imports. As noted above, these have significant energy penalties due to both the energy required for liquification and transportation, and to the capex needed to construct these facilities. The result is long-term contracts without any basis except oil for a reference price.

A consequence is that, unlike oil, there is no global natural gas market. The ability to shift a cargo of LNG from one port to another makes Asia essentially one market, but it is characterized by the high costs of LNG processing. The USA, with abundant new gas supplies coming on stream from tight gas formations and a developed pipeline network that only needs to support fairly short transportation distances, presently has low natural gas costs determined by spot prices. Europe is in an intermediate position. Both the IEA (2013) and the Russian Academy of Sciences (Mitrova 2013) predict that this fragmentation of the natural gas market will continue for decades. It might be noted that the Russians, who have very large reserves of natural gas, continue to insist, at least publicly, on long-term contracts with prices tied to the price of oil when negotiating new contracts.

Both oil and natural gas are used for their energy. But the inflexibility of the natural gas supply can be seen in the unregulated US market. Because so much gas is currently available, the energy cost of the two fuels has diverged, as shown in Fig. 8.2.

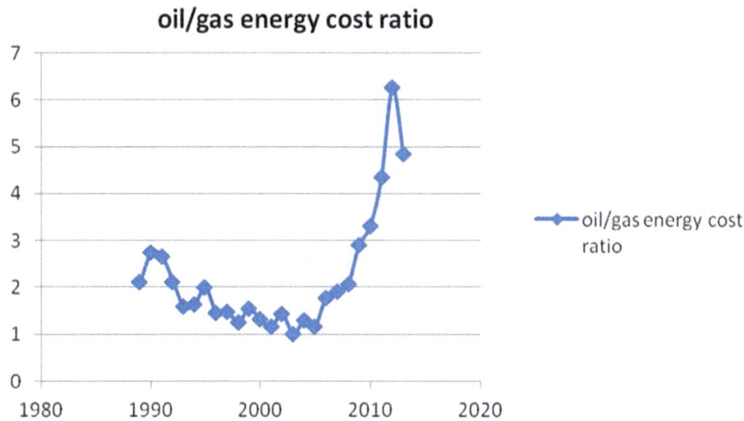

Fig. 8.2 Oil has tended to cost more than gas, due to its greater convenience in handling and storage. The rapid increase in oil prices after 2005 is clear

The fact that it is so easy to convert an internal combustion engine from oil to natural gas as discussed in Chap. 6, makes such conversions an economic inevitability when such price disparities exist. Whether the advantage will persist is another matter. Nevertheless, the tinkering with the technology combined with the large amounts of gas available world-wide, point to a continuation of oil replacement by natural gas.

Chapter 9
"Muddling Through"

If most of the previous chapters have seemed to lack a clear view of the future, you have read them correctly. The current world situation is fraught with uncertainty and tension. There are major open questions about oil availability. Problems related to sustainability, in particular global warming, become more visible each day. Both the global and the political situations of many specific countries are trending to instability and all of the above are inter-related. To highlight the uncertainties and confusions that relate particularly to the oil industry:

1. The world is finite and there is little more of the cheap, easy oil still to be found; remaining production of low-cost oil is concentrated in the Gulf area of the Middle East. But there is lots of oil yet in the ground, much of it unaccounted for by peak oil because it is outside the economic boundaries of that analysis. The questions are whether the oil still to be produced will be sufficiently cheap and whether it can be produced at the rates needed to provide sufficient energy to the global economy.
2. The amount of natural gas that is physically available is at least equally great in energy content as oil. Because there are additional exploration opportunities for natural gas, the amounts may be even greater than current reserve estimates. But the up-front capital investments required are at odds with the fast-changing economic and political environments that characterize human societies in the world today.

The original Hubbert peak applied only to conventional oil onshore in the contiguous USA, predicting maximum production for 1970. The global oil production peak was forecast for the first decade of the twenty-first century. Both these forecasts were essentially correct, with the 1970s and the 2000s being decades of increasing oil prices and consequent economic slowdowns. But the Hubbert oil analysis was only for traditional oil accumulations, and there is much uncertainty about how much additional oil may be available. One thing about this additional oil is certain, however: it will be more costly. Both in monetary and energy terms, the non-traditional supplies will cost more than what we have been used to paying.

S.W. Carmalt, *The Economics of Oil*, SpringerBriefs in Energy,
DOI 10.1007/978-3-319-47819-7_9

Economist Jay Forrester invented a game called the "Beer Distribution Game"[1] which illustrates how oscillations develop in an economic system. The basic framework is that beer retailers get orders from customers; retailers examine their stock and place orders with wholesalers; wholesalers examine their stock and place orders with distributors; and the distributors do the same with the brewers. Each player has costs for both holding unsold inventory and for running out of stock; costs which are asymmetric (running out costs more per case of beer than having excess inventory). The game also features time delays between placing an order and receipt of the beer ordered. Finally, the only communications between the players are the actual orders placed. Over many times of playing this game, the results show that the players always overreact to a perturbation in original customer orders, resulting in swings between over- and under-supply at levels in the system. The oil industry has some similarities. The long capital investment cycles needed for oil and gas production projects mean that along with all the other factors involved in projecting the future of the industry, some effects of oscillations should be added.

Commodity traders have a saying, "the cure for high prices is—high prices." The period of 1973–1985 was a period of high prices for oil. But by 1983, the new production from Alaska's North Slope, the North Sea, Mexico, and other non-OPEC projects was putting more and more oil on the market. For several years Saudi Arabia had lowered its production to keep the prices high, but eventually in 1985 the Saudis capitulated and lowered the price. In hindsight, this increase in production and thus downward pressure on prices should have been foreseen. But it was not, at least not generally within the oil industry. The decade of increasing prices had conditioned the attitudes of most people to the high prices with only increases in sight. The company I worked for was projecting $35/barrel oil in the early 1980s rising to the $70 or $80/barrel by the middle of the 1990s. Once the initial contracts are signed, many projects will go ahead, even in the face of slumping prices. This boom and bust cycle is what the Beer Game predicts.

The decline in oil prices since 2014 has some similarities to the collapse of oil prices in the 1980s. There is presently an over-supply of oil in the markets, and oil companies are struggling to keep up the profit expectations that arose in the decade before the decline. This means that many projects have been abandoned or deferred, which is likely to result in less oil being available, and therefore higher prices, sometime in the future. Oil "optimists" view all this as the predictable oscillations similar to the Beer Game, and argue that the world economy will adjust. But the fact that the resource must be finite, coupled with the fact that all the new oil sources have much lower EROIs, indicate that the days of inexpensive oil will soon be history. Oil "pessimists" think that the higher prices will be the forcing factor in profound global economic change.

In evaluating the oil industry for the coming decade, it is not only the finances of specific projects or even programs that determine how the economy will evolve; there are a number of non-financial issues. For one thing, if oil is considered as just

[1]The description of the Beer Distribution Game is taken from Beinhocker (2006, pp. 168–171).

one of the many minerals that are mined from the earth, the decline of EROI is analogous to ores that are less and less concentrated. For example, the copper ore being mined today has a grade of about 0.2%. That is, for every ton of rock mined, there is about one pound of copper and 1999 lb of rock that is of no economic value. The mining operation uses energy, first to extract the rock and then to do the separation. Much of the energy used is powered by diesel engines—trucks to haul the ore, crushers to pulverize the ore to allow separation, and so forth. Hence, as oil EROI is lowered, a copper mine is hit by two types of increasing costs: first is the increase in costs due to the grade of ore going down, and second is the increase in costs due to the oil becoming more scarce (Bardi 2014).

Copper is just one mineral. Everything on the earth is subject to the same double problem. Perhaps the two most important earth resources are agricultural land and fresh water. Most fresh water, Perkins (2012) says 92%, is used for agriculture. Without this, agricultural productivity decreases. Water can be pumped from one area to another. In California, 2–3% of total electricity is used just for one of the state's water projects; 90% of all electricity used on California farms is to pump water (Cohen et al. 2004). China is considering even more massive water transfer projects, to the considerable consternation of Pakistan, India, Myanmar, Thailand, Cambodia, and Vietnam which will be deprived of water that they now rely on. All these projects require energy, and if energy itself is a limited resource and therefore not available, it may be that resources such as water or fertile soil will prove to be the ones that constrain the economy.

Economists refer to factors outside of the normal framework of economic analysis as "externalities." In the case of oil and natural gas, the emission of CO_2 is an externality. Using the atmosphere as a free place to dispose of CO_2 is an externality, and the expense of global warming is not being paid by the emitters. The fact that the costs of a higher global temperature due to CO_2-induced warming are not included in the typical analysis of hydrocarbon projects does not mean that there are no economic costs; rather, it simply means that the costs have not been included. Various groups have attempted to put a figure on the value of each ton of CO_2 that is emitted into the atmosphere. If this cost becomes incorporated into the financial analyses of oil and natural gas companies, it will change the economics of the financial models used. Internally, the major companies do this as an exercise (Schapiro and Scorse 2014), but as yet this external cost is not applied to the real costs of a project.

Competition from other energy sources cannot be ignored, either. Photovoltaic cells (PV) for rooftops, wind turbines, solar thermal arrays, nuclear plants, hydroelectricity are all options to displace fossil fuels. Each of these has vocal advocates pointing out that the costs for at least some are falling rapidly, posing a direct economic challenge to fossil fuel energy. The cost benefits would likely be even greater with a charge for externalities, which is why many environmental economists propose some form of a carbon tax or charge. By charging for the "true cost" of fossil-fuel energy, the conversion to non-CO_2 emitting energy would become even more attractive. It should be noted that the present non-CO_2 energy sources use fossil-fuel energy in their own manufacture, which makes it more difficult to assess the true economic benefits of such changes.

It is possible to plot global population into the future. One can then plot GDP/person that one hopes to achieve, which is what leads to the GDP growth of India and China dominating most energy forecasts. It is possible to adjust these projections by making different assumptions about how the GDP/energy unit (roughly "efficiency") may change, particulary if one ignores Jevon's paradox.

Into this complex mixture, we must now add two more ingredients: finance and politics. The financial models that the oil industry uses are built around dollars. The use of dollars is an historical accident due primarily to the original US dominance of the oil industry. But what is money really?[2] Whatever the answer, one aspect of money is the ability to exchange value across both space and time. The financial future of oil industry decisions depends not only on whether the financial model is correct with respect to today's money, but whether it is correct with respect to future money.

Such interacting projections are the domain of systems analysis and systems theory. Researchers such as Randers (2012) have built models of great complexity in this way. But systems theory, even with its feedback loops, often does not capture entirely the fact that the interdependencies between different time series are extremely complex. The tools of multivariate analysis can be used to extract meaning from complex systems, and such analyses indicate that one of the challenges in predicting the economy is its multidimensional nature. The fossil fuels, fresh water, fertile soil, rare-earth metals, base metals, phosphorous, population, various pathogens, various species, and the list goes on, are each an axis in the analysis. Tverberg (2013) notes that Liebig's *Law of the Minimum*, which states that if any one critical input is lacking it will curtail the system despite plentitude in all other inputs, may be the controlling principle in oil. Tverberg suggests that energy, specifically oil, is the limiting input; it may also be the controlling principle in many of the other feedback loops of systems analysis. But the complexity of the system makes this difficult to prove and the limiting input may be something else.

Despite Darwin's initial insights, the exact mechanisms of evolution are still the subject of research. Of particular note is the development of the concept of "punctuated equilibrium" (Eldredge and Gould 1972), which proposes that a species may evolve very slowly over a long period of geologic time, and then a subgroup will evolve very quickly over a relatively short time to form a new species. Such punctuated equilibrium disruptions are observed in a number of other fields as well, and can be modeled by computer (see Beinhocker 2006, for discussions of avalanches and stock market examples).

At present, the world economic system is changing very rapidly. As outlined in Chap. 6, we are in a period of rapid change after the period of slow evolution from about 1945 to 2005. It is not peak oil alone that is driving this change, although the

[2]For an interesting discussion of the question of "what is money," which is fundamental to much of economics, see Mainelli and Harris (2011), pp. 222–225.

Table 9.1 Technologies and societal organization (after Jiska 2013)

Technology	Organization	Resource	Per capita income
Hunting	Tribes	Nature	100
Agriculture	Feudal	Farmland	1000
Industrial	National	Capital	10,000
Service	Global	Knowledge	100,000

high oil prices of the past decade are a contributor. Rather, it is that the economic and political "business-as-usual" approach is meeting diminishing returns on almost every front. The air in China's major cities has become unbreathable. A number of the world's major rivers are now so dammed that they provide no more water to their deltas and no more sediment to their productive coastal wetlands (Pearce 2006). The destruction of tropical rain forests continues, in particular in Brazil and Indonesia. And, accustomed to the cheap energy available since the start of the industrial revolution, the specific remedies proposed are frequently to put more energy into the system to keep the entropy down.

As this book goes to press, both the UK and the USA have populist-inspired political leaderships which are relatively hostile to taking political measures to address climate change. But politicians, despite periodic attempts to do so, cannot repeal basic natural laws. Once a fossil fuel is burned, it is no longer available as an energy source. On a finite planet, once used there will be less left for the future; and what is left will be, in most cases, more expensive.

Something is going to have to change. Jiska (2013) suggests that the human species has had four such changes, each of which allows the species to control an order of magnitude more resources, and hence have a better standard of living as shown in Table 9.1. The change from an Industrial to a Service economy is the economic change that is presently underway.

The problem with this view of a knowledge economy is the uncertainty as to whether knowledge is actually a raw material. Both hunting and agriculture are founded on direct solar energy; the industrial society is founded on fossil fuels. But what is the energy foundation of the service or information society? And what is its decreasing quality? Jiska has also observed that, over the very long term, the depletion of stored energy sources means that there is no choice (presuming the species survives) but to return to an essentially solar energy regime as illustrated in Fig. 9.1.

In a similar vein, the management guru Peter Drucker (1992) has proposed that the knowledge society will require a reorganization of social and political structures. Furthermore, knowledge alone is not sufficient. For several decades the best technocratic evidence has indicated that CO_2 emissions must be lowered. Naomi Oreskes (2014) suggests that the IPCC should abolish Working Group 1, the group that looks at the science of climate change. Her argument is that by not doing so, those that are opposed to dealing with the overwhelming consensus can continue to claim that the matter is still open to debate. Economists frequently take a different approach, arguing that even if the climate change projections are totally incorrect, the cost-benefit ratio of limiting CO_2 still represents a sound insurance policy at a

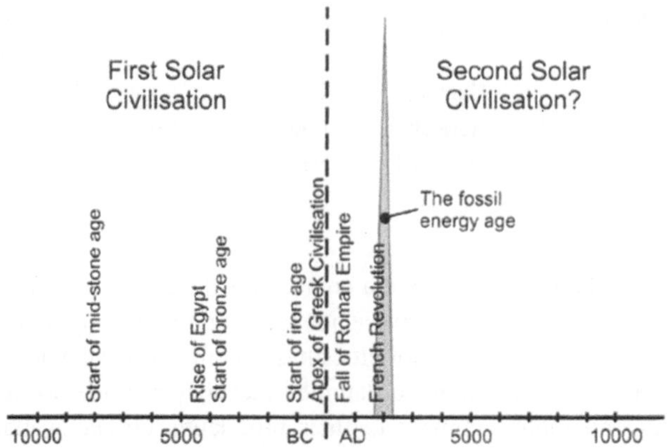

Fig. 9.1 Energy use of mankind (from Jiska 2013)

reasonable premium. Companies and individuals routinely purchase insurance; but as yet society as a whole is not prepared to move forward with any sort of climate change insurance. Both approaches point to the necessity of political rather than technical action.

As has been pointed out, one of the problems is the difference in time horizons between the financial investments being made and the structural changes that need to be made. This difference in time has a number of causes, but it is also clear that the financial horizon has been getting shorter. Companies no longer issue 99-year bonds. As economist Paul Krugman (2014) has pointed out, the long-term investments required for fundamental changes may require decisions outside the market capitalism framework.

Without a clear direction, and given capital constraints, some oil companies will make more opportune investments than others. In the past decade, BP has attempted and failed to become a partner with Russia in providing hydrocarbons to the global economy. Until recently Shell seemed to be skewing its capital budgets towards Arctic ventures. The very large companies have all participated in tight oil and gas plays. But there is no large company consensus as to what is going to be profitable. Meanwhile, tight plays are the frontier hydrocarbon resource with the lowest capital requirements, and hence the greatest ability to move uncertainty into controlled risk and eventual financial gain. The amounts of hydrocarbons eventually available from tight plays may not be as great as those available in harsh geographic environments or in areas of high political risk. But as was observed in Chap. 2, the goal of a company is to make money and only secondarily to provide oil or natural gas.

The above financial framework illustrates the sorts of uncertainties that the oil companies face in today's environment. But I believe that the political uncertainties will be even more important in determining what becomes of the oil industry over the coming two decades.

Until the Paris accords in December 2015, the effects of global warming and climate change had not been addressed on an international level. Some will argue that the Paris agreement is too weak to have changed this situation. But it is also clear that most people are becoming aware of climate change impacts. People are aware because the changes are costly. One response is for the government to fix the problem. As changes become more evident such demands will become more pressing. In 2011 I was ridiculed in the environmental NGO in which I was working at the time for predicting that, by the end of this decade there would be a loud demand in OECD countries for politicians to "fix" the climate. Such a loud cry may backfire by enabling self-aggrandizing demagogues, but more likely it will push the political system to increased intervention with respect to the use of fossil fuels.

The oil companies, in addition to the legal and financial frameworks within which they operate, also need a "social license" to engage in their business. The current investigation of ExxonMobil for fraud in connection with suppression of global warming studies is a small indication that this social license is being reconsidered. The fact that major oil and gas companies include some form of carbon charge in their internal economics indicates that they realize this threat, at least to some extent. But how this will evolve, and how much scope the political system will give these companies, is becoming a more evident issue.

If society, through its political processes, redirects the efforts of oil companies into new areas, then the companies will follow. But until such re-direction is provided by the political system, the oil companies will continue to do what they know how to do, which is to make as much profit as they can by providing oil and gas to the economy.

References

Adelman, M.A. 1993. The Economics of Petroleum Supply: Papers by M. A. Adelman, 1962–1993, 576 pp. Cambridge, MA: MIT Press.

Ahlbrandt, T.S. 2012. The Shift from the Static Hydrocarbon Model (Hubbert) to a Dynamic Model: Re-Evaluating the Hubbert Curve and the Global Petroleum Revolution: A New Era*, vol. 70116. Tulsa, Okla: Search and Discovery (AAPG), (2012), no. 13 February 2012. http://www.searchanddiscovery.com/documents/2012/70116ahlbrandt/ndx_ahlbrandt.pdf. Accessed October 26, 2014.

Aleklett, K. 2012. Peeking at Peak Oil, 325 pp. New York: Springer.

Alfaro, J.C., C. Corcoran, K. Davies, F.G. Pineda, G. Hampson, D. Hill, M. Howard, J. Kapoor, N. Moldoveanu, and E. Kragh. 2007. Reducing Exploration Risk. *Oilfield Review* 19(1): 26–43.

American Physical Society. 2014. Energy Units. http://www.aps.org/policy/reports/popa-reports/energy/units.cfm. Accessed September 12, 2014.

Andrews, S., and R. Udall. 2003. Oil Prophets: Looking at World Oil Studies Over Time. In 2nd International Workshop on Oil Depletion, 15 pp. Paris: Association for the Study of Peak Oil, May 2003.

Arkhipov, I., S. Bierman, and R. Chilcote. 2014. Russia says an Arctic well it drilled with Exxon Mobil has just struck oil—a lot of it. Toronto: Financial Post (quoting Bloomberg new reporting). http://business.financialpost.com/news/energy/russia-exxon-oil?__lsa=cdf1-481a. Accessed January 29, 2015.

Baer, Robert. 2003. The Fall of the House of Saud. The Atlantic (on-line edition). May 2003. http://www.theatlantic.com/magazine/archive/2003/05/the-fall-of-the-house-of-saud/304215/. Accessed October 28, 2014.

Baker Hughes. 2014. Rig Count Overview & Summary Count. http://phx.corporate-ir.net/phoenix.zhtml?c=79687&p=irol-rigcountsoverview. Accessed October 30, 2014.

Bardi, U. 2013. Cassandra's legacy: Peak shale oil? What peak? http://cassandralegacy.blogspot.it/2013/08/peak-shale-oil-what-peak.html. Accessed August 2, 2013.

Bardi, U. 2014. Extracted : How the quest for mineral wealth is plundering the planet, 247 pp. White River Junction, VT, USA: Chelsea Green Publishing.

BBC. 2014. South China Sea dispute. London. http://www.bbc.com/news. Accessed October 27, 2014.

Beinhocker, E.D. 2006. The Origin of Wealth, 454 p. Boston, MA: Harvard Business School Press.

Bentley, R. 2002. Oil forecasts past and present. *Energy Exploration and Exploitation* 20(6): 481–492.

Bentley, R.W. 2016. Introduction to Peak Oil. Lecture Notes in Energy, 104 pp. Cham (ZG), Switzerland: Springer.

Bentley, R., R. Miller, S. Wheeler, and G. Boyle. 2009. [TR 7] Comparison of global oil supply forecasts, Review UKERC/WP/TPA/2009/022. Review of Evidence on Global Oil Depletion, 86 pp. London: UK Energy Research Centre.

S.W. Carmalt, *The Economics of Oil*, SpringerBriefs in Energy,
DOI 10.1007/978-3-319-47819-7

Berman, A. 2010. Shale Gas—Abundance or Mirage? Why The Marcellus Shale Will Disappoint Expectations. http://www.theoildrum.com/node/7075. Accessed June 20, 2014.
Berman, A. 2015. Years Not Decades: Proven Reserves and the Shale Revolution. Houston, Texas: Houston Geological Society. Meeting 2015-02-23. https://www.youtube.com/watch?v=5Ae1fg44l7E. Accessed May 11, 2015.
Blundell, K. 2015. Black Holes—A very short introduction, 93 pp. Oxford: Oxford University Press.
BP. 2013. Statistical Review of World Energy 2013. http://www.bp.com/en/global/corporate/energy-economics/statistical-review-of-world-energy.html. Accessed October 21, 2014.
BP. 2014. Statistical Review of World Energy 2014. http://www.bp.com/en/global/corporate/energy-economics/statistical-review-of-world-energy.html. Accessed June 17, 2015.
BP. 2015. Statistical Review of World Energy 2015. https://www.bp.com/content/dam/bp/pdf/energy-economics/statistical-review-2015/bp-statistical-review-of-world-energy-2015-full-report.pdf. Accessed June 22, 2016.
BP. 2016. Statistical Review of World Energy 2016. http://www.bp.com/en/global/corporate/energy-economics/statistical-review-of-world-energy.html. Accessed July 27, 2016.
BP Statisticals (Year); *see* BP; (year); Statistical Review of World Energy (year).
Brandt, A.R., T. Yeskoo, and K. Vafi. 2015. Net energy analysis of Bakken crude oil production using a well-level engineering-based model. Energy 93(part 2):2191–2198. Doi:10.1016/j.energy.2015.10.113.
Business Week. 2013. In China, the License Plates Can Cost More Than the Car. http://www.businessweek.com/articles/2013-04-25/in-china-the-license-plates-can-cost-more-than-the-car. Accessed October 27, 2014.
Campbell, C.J. 1997. The Coming Oil Crisis, 210 pp. Bretwood CM14 4RX (UK): Multi-Science Publishing Company.
Campbell, C.J. 2013. Campbell's atlas of oil and gas depletion, 411 pp. New York: Springer.
Campbell, C.J., and J. Laherrere. 1998. The End of Cheap Oil. *Scientific American* 278(3): 78–83.
Capra, F., and P.L. Luisi. 2014. The Systems View of Life, 452 pp. Cambridge (UK): Cambridge University Press.
Carmalt, S.W., and B. St. John. 1986. Giant Oil and Gas Fields. In *Future Petroleum Provinces of the World*, vol. 40, ed. Halbouty, Michael, T., 11–53. Tulsa, OK, USA: AAPG. AAPG Memoir.
Carrington, D. 2013. Carbon bubble will plunge the world into another financial crisis. Manchester/London (UK): The Guardian. 2013-04-19. http://www.theguardian.com/environment/2013/apr/19/carbon-bubble-financial-crash-crisis. Accessed October 28, 2014.
Carroll, L. (alias for C. L. Dodgson). 1871. Through the Looking Glass. London: Macmillan.
Center for Energy Economics. 2006. How much does LNG cost. http://www.beg.utexas.edu/energyecon/lng/LNG_introduction_09.php. Accessed October 30, 2014.
Chandra, V. 2014. Liquefied Natural Gas (LNG), Industry Association: http://www.natgas.info/gas-information/what-is-natural-gas/lng. Accessed January 13, 2015.
Cleveland, C.J., R. Costanza, C.A.S. Hall, and R. Kaufmann. 1984. Energy and the U.S. Economy. A Biophysical Perspective. *Science* 225(4665): 890–897.
Cohen, R., B. Nelson, and G. Wolff. 2004. Energy Down the Drain, 78 pp. San Francisco and Oakland, CA: Natural Resources Defense Council.
Craig, J. 2014. How the Bakken Oil Boom Impacts U.S. Supply Chains. http://blog.chrobinson.com/oil-and-gas/bakken-oil-boom-impacts-u-s-supply-chains/. Accessed October 28, 2014.
Crooke, A. 2014a. Middle East Time Bomb: The Real Aim of ISIS Is to Replace the Saud Family as the New Emirs of Arabia. New York: Huffington Post. http://www.huffingtonpost.com/alastair-crooke/isis-aim-saudi-arabia_b_5748744.html. Accessed October 08, 2014.
Crooke, A. 2014b. You Can't Understand ISIS If You Don't Know the History of Wahhabism in Saudi Arabia. New York: Huffington Post. http://www.huffingtonpost.com/alastair-crooke/isis-wahhabism-saudi-arabia_b_5717157.html?utm_hp_ref=world. Accessed October 08, 2014.

Davenport, C. 2014. Emissions from India Will Increase. Official Says. New York: New York Times. http://www.nytimes.com/2014/09/25/world/asia/25climate.html?_r=1. Accessed September 26, 2014.
Deffeyes, K.S. 2001. Hubbert's Peak : The Impending World Oil Shortage, 190 pp. Princeton, NJ: Princeton University Press.
Deffeyes, K.S. 2005. Beyond Oil, 188 pp. New York: Hill and Wang (div. of Farrar, Straus and Giroux).
Demytrie, R. 2012. Development challenge of Kazakhstan's giant oilfield. London: BBC. http://www.bbc.com/news/world-asia-20251682. Accessed October 30, 2014.
Drucker, P.F. 1992. The New Society of Organizations. *Harvard Business Review* 70(5): 95–104.
Durham, L.S. 2016. U.S. Shale Production Proves Resilient to Price Collapse. *AAPG Explorer* 37 (7): 16.
EIA. 2013. Few transportation fuels surpass the energy densities of gasoline and diesel. http://www.eia.gov/todayinenergy/detail.cfm?id=9991. Accessed October 26, 2014.
EIA. 2014a. China—Country Energy Analysis. Washington, D.C., USA: EIA, 37 pp. http://www.eia.gov/countries/analysisbriefs/China/china.pdf. Accessed October 27, 2014.
EIA. 2014b. India—Country Energy Analysis. Washington, D.C., USA: EIA. 26 pp. http://www.eia.gov/countries/analysisbriefs/India/india.pdf. Accessed October 27, 2014.
EIA. 2016a. International Energy Outlook 2016. http://www.eia.gov/forecasts/ieo/. Accessed July 15, 2016.
EIA. 2016b. International Energy Statistics (a) Energy intensity—GDP/Energy—1980–2011 (b) CO_2 emissions 1980–2011 (c) Population 1980–2011 (d) Total Primary Energy consumption 1980–2011 (e) Total Primary Energy production 1980–2011. http://www.eia.gov/cfapps/ipdbproject/IEDIndex3.cfm?tid=3&pid=26&aid=24. Accessed July 06, 2016.
Eldredge, N., and S.J. Gould. 1972. Punctuated equilibria: an alternative to phyletic gradualism. In Models in Paleobiology, 82–115. San Francisco, CA: Freeman Cooper.
Engdahl, F.W. 2014. The Secret Stupid Saudi-US Deal on Syria. Oil Gas Pipeline War. Montreal, Quebec, Canada: Center for Research on Globalization. http://www.globalresearch.ca/the-secret-stupid-saudi-us-deal-on-syria/5410130. Accessed October 28, 2014.
ExxonMobile. 2012. Exxon Outlook to 2040, 54 pp. Irving, Texas: ExxonMobile Corp.
FASB. 2010. Extractive Activities—Oil and Gas (Topic 932) : Oil and Gas Reserve Estimation and Disclosures ASU 2010-XX Extractive Activities—Oil and Gas Topic 932 Oil and Gas Reserves: Norwalk, Conn., USA: Financial Accounting Standards Board.
Forbes, B., and H. Wilczynski. 2010. "Flexible factory" steadies unconventional gas work. Oil & Gas Journal 2010-03-15. http://www.ogj.com/articles/print/volume-108/issue-10/General-Interest/-flexible-factory.html. Accessed October 29, 2014.
Foss, M.M. 2007. Introduction to LNG. Austin, Texas: Bureau of Economic Geology. 40 pp. http://www.beg.utexas.edu/energyecon/lng/documents/CEE_INTRODUCTION_TO_LNG_FINAL.pdf. Accessed January 13, 2015.
Frantz, J.H., Jr. 2014. Tight shale plays : Pennsylvania. SPE Distinguished Lecture. Geneva, Switzerland. 2014-09-29.
Friedemann, A.J. 2016. When Trucks Stop Running: Energy and the Future of Transportation, Springer briefs in Energy, 130 pp. London: Springer.
Gallucci, V.F. 1973. On the principles of thermodynamics in ecology. In Annual Review of Ecology and Systematics, vol. 4, 329–357. Palo Alto, Annual Reviews.
Garcia, D. 2009. New World Model—EROEI issues. http://europe.theoildrum.com/node/5688. Accessed October 31, 2014.
Georgescu-Roegen, N. 1971. The Entropy Law and the Economic Process, 457 pp. Cambridge, MA: Harvard University Press.
Geuss, M. 2014. Porsche, Mercedes building electric cars to challenge Tesla: Ars Technica: Cambridge, MA. http://arstechnica.com/cars/2014/10/porsche-mercedes-building-electric-cars-to-challenge-tesla/. Accessed October 27 2014.

Gleick, J. 1988. “Slippery Water”: Mystery Seems Finally Solved. New York: New York Times. http://arstechnica.com/cars/2014/10/porsche-mercedes-building-electric-cars-to-challenge-tesla/. Accessed October 27, 2014.

Hall, C.A.S. 1972. Migration and metabolism in a temperate stream ecosystem. *Ecology* 53: 585–604.

Hall, C.A.S., and K.A. Klitgaard. 2012. Energy and the Wealth of Nations : Understanding the Biophysical Economy, 402 pp. New York: Springer.

Hall, C., M. Lavine, and J. Sloane. 1979. Efficiency of energy delivery systems: I. An economic and energy analysis. Environmental Management 3(6): 493–504. Doi:10.1007/BF01866318.

Hall, C., E. Kaufman, S. Walker, and D. Yen. 1979. Efficiency of energy delivery systems: II. Estimating energy costs of capital equipment. Environmental Management 3(6): 505–510. Doi:10.1007/BF01866319.

Hallock Jr., J.L., W. Wu, C.A.S. Hall, and M. Jefferson. 2014. Forecasting the limits to the availability and diversity of global conventional oil supply. *Energy* 64: 130–153. doi:10.1016/j.energy.2013.10.075.

Hanner, I. 2014. Gazprom and CNPC Sign Technical Agreement on Power of Siberia. Exploration World. http://explorationworld.com/pipelines/153/Gazprom-and-CNPC-Sign-Technical-Agreement-on-Power-of-Siberia. Accessed October 29, 2014.

Hardin, G. 1968. The Tragedy of the Commons. *Science* 162(3859): 1243–1248. Doi:10.1126/science.162.3859.1243.

Hargreaves, S. 2012. World’s 10 most expensive energy projects—#1—Kashagan—$116 billion. CNN. http://money.cnn.com/gallery/news/economy/2012/08/27/expensive-energy-projects/10.html. Accessed October 30, 2014.

Haun, J.D. (ed). 1975. Methods of Estimating the Volume of Undiscovered Oil and Gas Resources, vol. 1, 195 pp. Tulsa, OK, USA: AAPG. Studies in Geology.

Hayes, C. 2014. The New Abolitionism. The Nation. http://www.thenation.com/article/179461/new-abolitionism?page=full. Accessed October 28, 2014.

Helman, C. 2014. The World’s Biggest Oil Companies—In Photos. Forbes. http://www.forbes.com/pictures/mef45efkii/the-worlds-biggest-oil-companies-29/. Accessed December 11, 2014.

Henriksen, B.E. 2004. Basic petroleum economics. http://www.ccop.or.th/ppm/document/CHWS2/CHWS2DOC10_henriksen.pdf. Accessed October 31, 2014.

Hirsch, R. 2012. The Impending World Oil Shortage: Learning from the Past. ASPO Annual Meeting, Vienna, Austria. 30 May to 1 June, 2012. http://www.aspo2012.at/wp-content/uploads/2012/06/Hirsch_aspo2012.pdf. Accessed October 28, 2014.

Hirsch, R.L., R. Bezdek, and R. Wendling. 2005. *Peaking of World Oil Production: Impacts, Mitigation, & Risk Management*, 91. Washington, D.C., USA: US Department of Energy.

Horn, M.K. 2011. Giant Fields GIS Project. http://www.datapages.com/AssociatedWebsites/GISOpenFiles/HornGiantFields.aspx. Accessed October 23, 2014.

Hubbert, M.K. 1938. Determining The Most Probable. http://www.technocracy.org/technical-alliancetn/17-m-king-hubbert/205-hubbert. Accessed March 26, 2013.

Hubbert, M.K. 1956. Nuclear Energy and the Fossil Fuels. Houston, Texas: Shell Development Company. Publication No. 95. 40 pp. http://www.hubbertpeak.com/hubbert/1956/1956.pdf. Accessed July 16, 2012.

Hubbert, M.K. 1962. Energy Resources. Washington, D.C., USA: National Academy of Sciences, National Research Council Publication 1000-D, 141 pp.

Hubbert, M.K. 1967. Degree of Advancement of Petroleum Exploration in United States. *Bull AAPG* 51(11): 2207–2227.

Hubbert, M.K. 1969. Energy Resources. In Resources and Man. W. H. Freeman and Co.

Hubbert, M.K. 1974. U.S. Energy Resources: A Review as of 1972. U.S. Senate Committee 93–40 (92–75). Washington, D.C., USA, U.S. Senate.

Hubbert, M.K. 1982. Techniques of Prediction as Applied to the Production of Oil and Gas. In *Oil and Gas Supply Modelling*, ed. S.I. Gass, 16–141. Washington, D.C., USA: National Bureau of Standards (USA). Special Publication.

Hughes, J.D. 2014. Drilling Deeper—A Reality Check on U.S. Government Forecasts For a Lasting Tight Oil & Shale Gas Boom, 315 pp. Santa Rosa, CA, USA: Post Carbon Institute.

Hussain, Y. 2013. Oil shipping by rail: tank car demand rises at breakneck speed. http://business.financialpost.com/2013/02/22/demand-for-tank-cars-to-ship-crude-oil-by-rail-rises-at-breakneck-speed/?__lsa=6465-20d7. Accessed October 28, 2014.

IEA. 2012. Iraq Energy Outlook. http://www.iea.org/publications/freepublications/publication/WEO_2012_Iraq_Energy_OutlookFINAL.pdf. Accessed October 28, 2014.

IEA. 2013. World Energy Outlook 2013, 708 pp. Paris: International Energy Agency.

IEA. 2014. World Energy Outlook 2014, 600 pp. Paris: International Energy Agency.

Illinois State Geological Survey. 2012. Seismic Reflection. http://crystal.isgs.uiuc.edu/sections/geophys/seisref.shtml. Accessed October 31, 2014.

Inman, M. 2016. The Oracle of Oil : a maverick geologist's quest for a sustainable future, 337 pp. New York: W.W.Norton & Co., Inc.

International Energy Agency; *see* IEA.

Irfan, U. 2014. Burden of Germany's shift to renewable energy falls on taxpayers, but energy rates are close to U.S. range. Washington, D.C., USA: Climate Wire. https://energyindemand.com/2014/10/24/comparing-the-us-to-germanys-energy-transition/. Accessed September 11, 2016.

IRS. 2015. Publication 535 (2015), Business Expenses. Washington, D.C., USA: Internal Revenue Service.

Itar-Tass. 2014. Cost of South Stream gas pipeline off-shore section can climb over €10 billion. http://en.itar-tass.com/economy/753616. Accessed October 30, 2014.

Jevons, W.S. 1865. The Coal Question, 349 pp. London: Macmillan.

Jevons, W.S. 1871. The Theory of Political Economy, 294 pp. London/New York: Macmillan and Co.

Jischa, M.F. 2004. Studium der Umweltwissenschaften, 276 pp. Berlin: Springer.

Jischa, M.F. 2013. About climate change and future energy systems. *Swiss Bulletin für angewandte Geologie* 18(2): 83–93.

Kearney, A.T. 2013. GSMA Mobile Economy 2013. http://www.gsmamobileeconomy.com/GSMA%20Mobile%20Economy%202013.pdf. Accessed August 28, 2014.

Kilian, L., and R.J. Vigfusson. 2014. The Role of Oil Price Shocks in Causing U.S. Recessions. International Finance Discussion Paper 1114, 41 pp. Washington, D.C., USA: Federal Reserve Board.

King, C.W., C.A.S. Hall, 2011. Relating Financial and Energy Return on Investment. Sustainability 3(10): 810–1832. Doi:10.3390/su3101810.

Knight, F.H. 1921. *Risk, uncertainty, and profit*. Boston, MA: Houghton Mifflin & Co.

Knittel, C.R. 2011. Automobiles on Steroids : Product Attribute Trade-Offs and Technological Progress in the Automobile Sector. American Economic Review 2012. Doi:10.1257/aer.101.7.3368

Kopits, S. 2014. Citi vs. Chevron: two opposing views of the oil price future. http://blogs.platts.com/2014/04/09/citi-chevron-oil-price/. Accessed July 14, 2014.

Korchemkin, M. 2014. Gazprom claims "Power of Siberia" is one of its least expensive projects. Natural Gas Europe. http://www.naturalgaseurope.com/gazprom-power-of-siberia-pipeline-cost. Accessed October 29, 2014.

Kotzot, H., C. Durr, D. Coyle, and C. Caswell. 2007. LNG Liquefaction—Not all plants are created equal. Houston, Texas: KBR Article PS4-1. KBR (ex. Kellog, Brown & Root). 20 pp. http://www.kbr.com/newsroom/publications/technical-papers/lng-liquefaction-not-all-plants-are-created-equal.pdf. Accessed October 15, 2014.

Krugman, P. 2014. Ideology and Investment. New York: New York Times. http://www.nytimes.com/2014/10/27/opinion/paul-krugman-ideology-and-investment.html?module=Search&mabReward=relbias%3Ar%2C%7B%221%22%3A%22RI%3A7%22%7D&_r=0. Accessed October 27, 2014.

Kuykendall, C. 2005. Hubbert Bibliography Compilation Project. http://www.hubbertpeak.com/hubbert/bibliography.htm. Accessed September 02, 2014.

Larsen, D. 2014. ChemWiki: The Dynamic Chemistry E-textbook. E-textbook. http://chemwiki.ucdavis.edu/. Accessed October 24, 2014.

Leaton, J. 2012. Unburnable Carbon—Are the world's financial markets carrying a carbon bubble? 36 pp. London: Carbon Tracker Initiative.

Lifton, R.J. 2014. The Climate Swerve. New York: New York Times http://www.nytimes.com/2014/08/24/opinion/sunday/the-climate-swerve.html. Accessed October 28, 2014.

Likvern, R. 2012. Is Shale Oil Production from Bakken Headed for a Run with "The Red Queen"? http://www.theoildrum.com/node/9506. Accessed January 28, 2015.

Los Angeles Times. 2016. At 15 cents a gallon, it's the cheapest gas in the world—yet Venezuela worries. Los Angeles: Los Angeles Times. http://www.latimes.com/world/mexico-americas/la-fg-venezuela-gasoline-fears-20160219-story.html. Accessed July 20, 2016.

Mainelli, M., and I. Harris. 2011. The Price of Fish, 310 pp. London/Boston: Nicholas Brealey Publishing.

Manne, R. 2012. How vested interests defeated climate science: A Dark Victory. The Monthly (Australia). http://www.themonthly.com.au/issue/2012/august/1344299325/robert-manne/dark-victory. Accessed October 28, 2014.

Martinot, E., C. Dienst, L. Weiliang, and C. Qimin. 2007. Renewable Energy Futures : Scenarios and pathways. http://www.martinot.info/Martinot_et_al_AR32_prepub.pdf. Accessed December 09, 2014.

Maschhoff, B. 2013. Of Milk Cows and Saudi Arabia. http://www.theoildrum.com/node/10238. Accessed October 28, 2014.

McCabe, P.J. 1998. Energy Resources—Cornucopia or Empty Barrel? *AAPG Bulletin* 82(11): 2110–2134.

McGrath, M. 2013. Fracking: Water concerns persist? London: BBC. http://www.bbc.com/news/science-environment-23724657. Accessed October 30, 2014.

Meadows, D.H., D.L. Meadows, J. Randers, and W.W.I. Behrens. 1972. *The Limits to Growth*. New York: Universal Books.

Mearns, E. 2006. ASPO-USA: Support for Global Energy Flow modelling and a Net Energy database. http://www.theoildrum.com/story/2006/10/31/144929/65. Accessed October 02, 2015.

Mearns, E. 2013. Marcellus shale gas Bradford Co Pennsylvania: production history and declines. http://euanmearns.com/marcellus-shale-gas-bradford-co-pennsylvania-production-history-and-declines/. Accessed January 06, 2015.

Mearns, E. 2016. ERoEI for Beginners. http://euanmearns.com/marcellus-shale-gas-bradford-co-pennsylvania-production-history-and-declines/. Accessed May 26, 2016.

Michel, J. 2014. Can Germany survive the Energiewende? Energy Post. http://euanmearns.com/marcellus-shale-gas-bradford-co-pennsylvania-production-history-and-declines/. Accessed October 28, 2014.

Mitrova, T.A. 2013. Global and Russian energy outlook up to 2040. Moscow: Russian Academy of Sciences. 175 pp. https://www.eriras.ru/files/2014/forecast_2040_en.pdf. Accessed September 11, 2016.

Moscariello, A. 2016. Reservoir geo-modelling and uncertainty management in the context of geo-energy projects. *Swiss Bulletin für angewandte Geologie* 16(1): 29–43.

Murphy, D. 2011. The Energy Return on Investment Threshold. http://www.theoildrum.com/node/8625. Accessed September 20, 2014.

Murphy, D.J., and C.A.S. Hall. 2010. Year in review—EROI or energy return on (energy) invested. *Annals of the New York Academy of Science* 1185(2010): 102–118.

NASDAQ. 2015. Natural Gas Price: Latest Price & Chart for Natural Gas. http://www.nasdaq.com/markets/natural-gas.aspx. Accessed January 27, 2015.

Nave, R. 2000. Entropy. http://hyperphysics.phy-astr.gsu.edu/hbase/therm/entrop.html. Accessed October 24, 2014.

Nelder, C. 2012. What EROI tells us about ROI. http://peakoil.com/business/chris-nelder-what-eroi-tells-us-about-roi. Accessed October 25, 2014.

Nordhaus, W. 2014. William Nordhaus: A new model for climate treaties. http://blog.iiasa.ac.at/2014/06/24/william-nordhaus-a-new-model-for-climate-treaties/. Accessed October 13, 2014.
Odum, E.P., and G.W. Barrett. 2005. Fundamentals of Ecology. Victoria: Thomson Brooks/Cole.
Offshore Technology. 2014. The world's biggest offshore gas projects—Offshore Technology. http://www.offshore-technology.com/features/featurethe-worlds-biggest-offshore-gas-projects-4177223/. Accessed October 30 2014.
Ore, O. 1960. Pascal and the invention of probability theory. *American Mathematical Monthly* 67: 409–419.
Oreskes, N. 2014. Naomi Oreskes Imagines the Future History of Climate Change. New York: New York Times. http://www.nytimes.com/2014/10/28/science/naomi-oreskes-imagines-the-future-history-of-climate-change.html?emc=edit_th_20141028&nl=todaysheadlines&nlid=32693583&_r=1. Accessed October 28, 2014.
Oreskes, N., and E.M. Conway. 2010. Merchants of Doubt : How a Handful of Scientists Obscured the Truth on Issues from Tobacco Smoke to Global Warming, 355 p. New York: Bloomsbury Press.
Oweiss, I. 2014. Pricing of Oil. http://faculty.georgetown.edu/imo3/petrod/poo.htm. Accessed October 30, 2014.
Owen, D. 2010. The Efficiency Dilemma. *The New Yorker Dec* 20(2010): 78–79.
Ozimek, A. 2014. Note to the Skeptics: the World Will Change for Self-Driving Cars. Forbes. http://www.forbes.com/sites/modeledbehavior/2014/10/25/note-to-the-skeptics/. Accessed October 27, 2014.
Pan, I. 2013. Most operators are seeing declining well costs in the Bakken. New York: Market Realist. Overview of the Bakken Shale oil play 2013-12-12. http://marketrealist.com/2013/12/operators-seeing-declining-well-costs-bakken/. Accessed October 29, 2014.
Patterson, R. 2014. Post Carbon Institute's LTO Reality Check. http://peakoilbarrel.com/post-carbon-institutes-lto-reality-check/ Accessed October 29, 2014.
Pearce, F. 2006. When the rivers run dry : water, the defining crisis of the twenty-first century, 324 p. Boston, MA: Beacon Press.
Peet, N.J., and J.T. Baines. 1986. Energy analysis : a review of theory and applications, 56 pp. Auckland: New Zealand Energy Research and Development Committee.
PennEnergy. 2013. Seismic survey suggests oil and gas potential for Swala. http://www.pennenergy.com/articles/pennenergy/2013/10/seismic-survey-suggests-oil-and-gas-potential-for-swala.html. Accessed October 31, 2014.
Perkins, S. 2012. Is agriculture sucking fresh water dry? Science. http://news.sciencemag.org/2012/02/agriculture-sucking-fresh-water-dry. Accessed October 30, 2014.
Perlez, J. 2014. China and Russia Reach 30-Year Gas Deal. New York: New York Times. http://www.nytimes.com/2014/05/22/world/asia/china-russia-gas-deal.html?_r=0. Accessed October 30, 2014.
Petrostrategies. 2012. Drilling Success Rates: http://www.petrostrategies.org/Graphs/drilling_success_rates.htm. Accessed October 30, 2014.
Quick, A.N., and N.F. Buck. 1983. Strategic Planning for Exploration Management, 111 pp. Boston, MA: International Human Resources Development Corp.
Rabbitt, M.C. 2000. Establishment of the U.S. Geological Survey. U.S. Geological Survey Circular 1050. Washington, D.C., USA: U.S. Geological Survey.
Randers, J. 2012. 2052—A Global Forecast for the Next Forty Years, 351 pp. White River Junction, VT, USA: Chelsea Green Publishing.
Reinvang, R., and G. Peters. 2008. Norwegian Consumption, Chinese pollution : An example of how OECD imports generate CO_2 emissions in developing countries, 34 pp. Oslo: WWF Norway.
Rodrigue, J.-P. 2013. Supply, Demand and Equilibrium Price. E-textbook. https://people.hofstra.edu/geotrans/eng/methods/supplydemandprice.html. Accessed October 24, 2014.
Royal Dutch Shell. 2009. Shell energy scenarios to 2050, 52 pp. Den Haag: Shell International.

Rumsfeld, D.H. 2002. DoD News Briefing of February 12, 2002. Washington, D.C., USA. http://archive.defense.gov/Transcripts/Transcript.aspx?TranscriptID=2636. Accessed September 09, 2016.

Said, S. 2013. The Mystery of Saudi's Spare Oil Production Capacity. New York: Wall Street Journal. http://blogs.wsj.com/middleeast/2013/09/18/the-mystery-of-saudis-spare-oil-production-capacity/. Accessed October 09, 2014.

Sampson, A. 1975. The Seven Sisters : The great oil companies & the world they shaped, 334 pp. New York: Viking Press.

Schapiro, M., and J. Scorse. 2014. Oil Companies Quietly Prepare for a Future of Carbon Pricing. Yale Environment 360. http://e360.yale.edu/feature/oil_companies_quietly_prepare_for_a_future_of_carbon_pricing/2807/. Accessed September 24, 2014.

Scheck, J. 2013. A $30 Billion Hole in the Caspian Sea? New York: Wall Street Journal. http://online.wsj.com/articles/SB10001424127887324050304578412760496098192. Accessed October 29, 2014.

Sims, R.E.H., R.N. Schock, A. Adegbululgbe, J. Fenhann, I. Konstantinaviciute, W. Moomaw, H. B. Nimir, and B. Schlamadinger. 2007. Contribution of Working Group III to the Fourth Assessment Report of the Intergovernmental Panel on Climate Change, 2007—Chapter 4. Energy Supplies. Cambridge (UK): Cambridge University Press.

Skinner, B.J. 1976. A Second Iron Age Ahead? *American Scientist* 64(3): 258–269.

Skinner, B.J. 1986. Earth Resources, 166 pp. Englewood Cliffs, NJ, USA: Prentice-Hall. Prentice-Hall Foundations of Earth Sciences.

Smith, C.H. 2012. Why Energy May Be Abundant but Not Cheap. http://www.financialsense.com/contributors/charles-hugh-smith/why-energy-may-be-abundant-but-not-cheap. Accessed October 25, 2014.

SPE. 2011. PRMS Guidelines Nov 2011, 222 pp. Richardson, Texas USA: Society of Petroleum Engineers.

Tainter, J.A. 1988. The collapse of complex societies, 250 pp. Cambridge (UK): Cambridge University Press.

Tainter, J.A. 2010. Energy, complexity, and sustainability: A historical perspective. *Environmental Innovations and Societal Transitions* 1(1): 89–95. doi:10.1016/j.eist.2010.12.001.

Tanaka, N. 2009. Our Global Energy Future—Looking beyond the economic crisis. Warsaw.

Tu, K. 2013. Status and Prospects of the Chinese Coal Value. Edinburgh: Global Energy Systems Conference. June, 2013.

Tverberg, G. 2013. Our Oil Problems are Not Over! http://ourfiniteworld.com/2013/10/02/our-oil-problems-are-not-over/. Accessed October 31, 2014.

UNDESA. 2012. World Population Prospects: The 2012 Revision. New York: United Nations.

United Nations. 2015. World Population Prospects, the 2015 Revision. https://esa.un.org/unpd/wpp/Download/Standard/Population/. Accessed August 30, 2016.

US Department of Commerce, BEA. 2016. http://bea.gov/newsreleases/national/GDP/GDPnewsrelease.htm. Accessed September 09, 2016.

US Department of Energy. 2011. Comparing Energy Costs per Mile for Electric and Gasoline-Fuelled Vehicles, 1 p. Washington, D.C., USA: US Department of Energy.

US Department of Energy; Energy Information Agency; *see* entries under EIA.

USGS. 2000. USGS World Petroleum Assessment 2000 : description and results by USGS World Energy Assessment Team, Digital Data DDS-60. Reston, VA: US Geological Survey. USGS Digital Data Series, four CD-ROM set.

Wagner, A. 2013. International Fuel Prices : 2012–2013/Data preview April 2013, Power point GIZ2013-en-ifp2013. Bonn. Deutsche Gesellschaft für Internationale Zusammenarbeit (GIZ) GmbH. 7 p.

Walras, L.1874. Eléments d'économie politique pur ou théorie de la richesse sociale; Lausanne//Paris; L. Corbaz & Cie//Guillaumin & Cie. 407 p.

Warner, N.R., C.A. Christie, R.B. Jackson, and A. Vengosh. 2013. Impacts of Shale Gas Wastewater Disposal on Water Quality in Western Pennsylvania. Environmental Science and Technology, 47(20): 11849–11857. Doi:10.1021/es402165b.

Washington's Blog. 2014. Shale Fracking Is a "Ponzi Scheme" … "This Decade's Version of The Dotcom Bubble" … "A Lot In Common With the Subprime Mortgage Market Just Before It Melted Down". http://www.washingtonsblog.com/2014/09/shale-fracking-ponzi-scheme.html. Accessed October 29, 2014.

Wehrey, F., and K. Sadjadpour. 2014. Elusive Equilibrium: America, Iran, and Saudi Arabia in a Changing Middle East. E-journal. http://carnegieendowment.org/2014/05/22/elusive-equilibrium-america-iran-and-saudi-arabia-in-changing-middle-east. Accessed October 28, 2014.

Welch, C. 2014. Hydraulic Fracturing. http://www.earthlyissues.com/gasdrill.htm. Accessed October 30, 2014.

White, D. 1919. Unmined supply of petroleum in the United States. *Journal Society of Petroleum Engineers* 12(5): 361–363.

Wilson, J.S., T. Otsuki, and M. Sewadeh. 2002. Dirty exports and environmental regulation : do standards matter to trade? 34 pp. Washington, D.C., USA: World Bank.

Wong, E. 2014. Most Chinese Cities Fail Minimum Air Quality Standards, Study Says. New York: New York Times. http://www.earthlyissues.com/gasdrill.htm. Accessed October 27, 2014.

World Bank. 2012. Global Gas Flaring Reduction—Estimated Flared Volumes from Satellite Data, 2007–2011. Washington, D.C., USA: World Bank. http://web.worldbank.org/WBSITE/EXTERNAL/TOPICS/EXTOGMC/EXTGGFR/0,,contentMDK:22137498~menuPK:3077311~pagePK:64168445~piPK:64168309~theSitePK:578069,00.html. Accessed October 30, 2014.

World Bank. 2015. GDP (constant 2005 US$) | Data | Table. http://data.worldbank.org/indicator/NY.GDP.MKTP.KD/countries?display=default. Accessed May 13, 2015.

World Bank. 2016. Data—GDP at market prices (current US$). Washington, DC. http://data.worldbank.org/indicator/NY.GDP.MKTP.CD. Accessed July 06, 2016.

Yergin, D. 1991. The Prize, 781 pp. New York: Simon & Schuster. (+ tables).

Index

S.W. Carmalt, *The Economics of Oil*, SpringerBriefs in Energy,
DOI 10.1007/978-3-319-47819-7